Pesticide Effects on Soil Microflora

Pesticide Effects on Soil Microflora

Edited by

L. Somerville
and
M.P. Greaves

Taylor & Francis
London, New York, Philadelphia
1987

UK   Taylor & Francis Ltd, 4 John St, London WC1N 2ET

USA   Taylor & Francis Inc., 242 Cherry St, Philadelphia,
           PA 19106-1906

British Library Cataloguing in Publication Data

Pesticide effects on soil microflora.
  1. Pesticides——Environmental aspects
  2. Micro-organisms——Effect of pesticides
on
  I. Somerville, L. II. Greaves, Michael P.
  632'.95042   QR100

ISBN 0-85066-365-2

*Printed in Great Britain by Taylor & Francis (Printers) Ltd,
Basingstoke, Hants.*

CONTENTS

PREFACE

Pesticides are an essential component of modern farming practice. By their extensive use developed nations have been able to turn to land-saving intensive agriculture to produce their total food needs. Parallel benefits in food quality have also been obtained while the farmer has benefited from reductions in labour associated with mechanical weed control.

While there is generally no shortage of land or labour for food cultivation in developing nations, much of their food deficit is attributable to pest damage either pre- or post-harvest. Although pesticides are not quite as extensively used in these countries at the present time, chemical pest control is becoming much more important as agricultural production is modernised.

The need for pesticides, therefore, is clearly established but it should be borne in mind that their discovery and use forms part of recent history. Although some inorganics such as "Bordeaux Mixture" were available prior to the 20th Century the "synthetic organic" pesticides originate only from World War II and the discovery of DDT. During the past 40 years the pesticide industry has flourished and in 1987 approximately 600 active ingredients are available to the world's farmers. These active ingredients in turn give rise to thousands of individual products some of which are manufactured in thousands of tonnes per annum. Indeed in 1986 total pesticide use world wide was estimated at 3.5 million tonnes.

Pesticides by definition are biologically active compounds. It follows, therefore, that they have the potential to be active in any sector of the environment in which they occur. While a farmer deliberately sprays his crop or livestock to achieve pest control, inevitably some part of the spray application misses the target site and reaches the soil. This text reviews the effect of this indirect and unintentional introduction of the pesticide to the soil microbial ecosystem and considers whether or not long-term environmental damage is likely to occur.

Reviews are included from leading scientists in industry, academia and government regulatory authorities. While pesticides are extensively regulated, the validity of current requirements is considered and modifications are suggested. Fundamental aspects such as the selection, collection and storage of soils are discussed since a "viable" soil sample is essential for any experimental evaluation. Thereafter the effects of pesticides on microbial systems such as soil respiration and nitrogen transformation are evaluated in "ring tests" undertaken within the European agrochemical industry. Modification of the test systems are proposed as well as alternative procedures. The regulatory viewpoint is also included.

The editors acknowledge the helpful cooperation of the authors and the extensive facilities made available by Schering Agrochemicals Ltd.

Special thanks are due to Sue Robins for preparing the text with consummate skill, to Louise McGlue for secretarial help and to Laurence Somerville Jr. for assisting in indexing the text.

Laurence Somerville,
Chesterford Park,
April, 1987.

CONTRIBUTORS

J.P.E. Anderson,
Bayer Ag, PF-A-CE,
Institut für Ökobiologie,
Pflanzenschutzzentrum,
Monheim D5090 Leverkusen.
F.R. of Germany.

D.J. Arnold,
Schering Agrochemicals Limited,
Environmental Sciences Department,
Chesterford Park Research Station,
Saffron Walden,
Essex. CB10 1XL.
U.K.

D. Barug,
TNO,
P.O. Box 108,
3700 AC Zeist,
The Netherlands.

K.A. Cook,
Shell Research Ltd.,
Sittingbourne Research Centre,
Sittingbourne,
Kent. ME9 8AG.
U.K.

K. De Coninck,
Laboratory of Soil Fertility
and Biology,
Kard, Mercierlaan 92,
B 3030 Leuven,
Belgium.

K.H. Domsch,
Bundesforschungsanstalt für Landwirtschaft
Bundesallee 50,
D 3300 Braunschweig,
F.R. of Germany.

CONTRIBUTORS

R. Dressen,
Laboratory of Soil Fertility
and Biology,
Kard, Mercierlaan 92,
B 3030 Leuven,
Belgium.

J-C. Fournier,
INRA,
17 Rue Sully,
21034 Dijon Cedex,
France.

M.P. Greaves,
AFRC Institute of Arable Crops Research,
Long Ashton Research Station,
Long Ashton,
Bristol,   B518 9AF.
U.K.

R.T. Hamm,
BASF Aktienfessellschaft,
Landw. Versuchsstation,
Postfach 220,
DO6703 Limburgerhof,
F.R. of Germany.

S. Horemans,
Laboratory of Soil Fertility
and Biology,
Kard, Mercierlaan 92,
B 3030 Leuven,
Belgium.

C.R. Leake,
Schering Agrochemicals Limited,
Environmental Sciences Department,
Chesterford Park Research Station,
Saffron Walden,
Essex.   CB10 1XL.
U.K.

CONTRIBUTORS

H-P. Malkomes,
Biologische Bundesanstalt,
für Land-und Forstwirtschaft,
Messeweg 11/12,
D-3300 Brauschweig,
F.R. of Germany.

H. Schuepp,
Swiss Federal Research Station,
CH-8820 Wadenswil,
Switzerland.

L. Somerville,
Schering Agrochemical Limited,
Environmental Sciences Department,
Chesterford Park Research Station,
Saffron Walden,
Essex. CB10 1XL.
U.K.

G. Soulas,
INRA,
27 Rue Sully,
21034 Dijon Cedex,
France.

N. Taubel,
Hoechst Ag,
Reisebuere C 820,
D6230 Frankfurt 80,
F.R. of Germany.

H. van Dijk,
Inst. for Soil Fertility,
P.O. Box 3003,
9750, RA Haren Gn,
The Netherlands.

A.M. van Doorn,
Institute for Pesticide Research,
Maykeweg 22,
6709 Pg Wageningen,
The Netherlands.

CONTRIBUTORS

H. Van der Werf,
Lab. Voor Alg. en Ind. Microbiologie,
Fac. Landb. wetensch,
R.U.G. Coupure 533,
B 9000 Gent,
Belgium.

W. Verstraete,
Lab. Voor Alg. en Ind. Microbiologie,
Fac. Landb. wetensch,
Landwirtschaft,
R.U.G. Coupure 533,
B 9000 Gent,
Belgium.

K. Vlassak,
Laboratory of Soil Fertility
and Biology,
Kard, Mercierlaan 92,
B 3030 Leuven,
Belgium.

J. Vonk,
TNO,
P.O. Box 108,
3700 AC Zeist,
The Netherlands.

INTRODUCTION

K. H. Domsch and M. P. Greaves

The ever-increasing demands for world agriculture to produce higher yields of better quality produce have created commensurate pressures to combat the losses caused by pests, weeds and diseases. The agrochemical industry has met this challenge most effectively and produced very rapidly a vast array of highly active chemicals for use in most crops in the world. In the majority of instances, in the developed countries at least, each crop may receive one or more application of several pesticides during each season. The benefits of such inputs are clearly visible in the form of greatly increased yields, so much so that overproduction is currently a problem in some areas. Unfortunately, the benefits may be accompanied by disadvantages. Pesticides are designed to be biologically active and, while every effort is made to ensure that this activity is confined to the intended target, absolute specificity has not been achieved and non-target organisms, including man, may be at risk.

Recognition of the risks attendant on pesticide usage has led many countries to develop schemes to regulate their sale and use. These schemes were initially concerned principally with factors such as toxicity towards mammals and birds, the obvious major organisms at risk. As public awareness of the potential hazards of pesticide use was heightened by publicity campaigns during the 1960's, so regulatory schemes were extended to cover the newly perceived risks. At this time, concern developed strongly that continuous inputs of pesticides to soils might affect the soil microflora and so impair soil fertility. Consequently, requirements were placed on pesticide manufacturers, notably by the

U.S. Environmental Protection Agency, to provide data for their pesticides that would enable some prediction of their likely effects on soil microflora and their activities. Perhaps because of the haste with which the requirements were developed, or due to too much emphasis on legislative aspects, these early efforts were characterized by an evident lack of scientific validity.

In the early 1970's, there were indications that regulatory authorities in Europe were considering the introduction of the requirement for tests of pesticides' effects on soil microflora. Previous experience of similar requirements stimulated scientists involved in pesticide research in the Federal Republic of Germany to consider and identify the most appropriate test methods that were available to meet this requirement. It was hoped that a consolidated view expressed by scientists to the regulatory authorities might influence the formulation of required test protocols. In this way the emphasis might be placed properly on scientific validity rather than on political or legislative aspects. A series of four symposia were organised jointly by the Biologische Bundesanstalt and the Bundesforschungsanstalt für Landwirtschaft at Braunschweig during the period 1973-77. The aim of these meetings was to bring together scientists, manufacturers and regulatory authorities to discuss the present knowledge of means of testing and assessing side-effects of pesticides on the soil microflora. It was hoped to produce an agreement as to the most meaningful tests and, where necessary, to institute simultaneous testing of methods in several laboratories.

While these meetings were highly successful, it was realized that it was necessary to extend the participation to a wider audience in order to reinforce the weight of any agreements reached. Accordingly, an international workshop was held at Braunschweig in 1978 and this was followed by a second in England in 1980. The aim of this latter meeting was to produce agreed recommendations on testing side-effects on soil microflora which were to be published and circulated as widely as possible. It was felt that, in this way, some positive influence might be brought to bear on both those authorities which had already imposed requirements for these tests and those who were in the process of developing them. The success of this meeting can be judged by the fact that within a very short time of the recommendations being published, they had been accepted as the basis of test requirements by several national regulatory authorities. In all, over a thousand copies of the recommendations were circulated throughout the world.

In the ensuing years, it has become increasingly apparent that, while the recommendations went a long way to meeting the needs, there were some inherent faults in the methods. That being so, it was equally clear that it was essential to reconsider them and publish a revised version. In order to achieve this revision it was decided to hold a further workshop, at Cambridge, in 1985. The contributors to that meeting were invited to prepare chapters for this book and the recommendations as to a revised test protocol are given in Appendix B. As before, these recommendations have been published separately and circulated as widely as possible, particularly to regulatory authorities and those scientists in government and industrial research who are directly concerned with pesticide effects on the soil environment and its inhabitants.

It is an inherent part of a scientists' make up, to disagree with the findings of his colleagues, or so it would appear from the contradictions present in much of the published literature. Indeed, this was also evident in the lively discussions following the presentations to the workshop. It is a great credit to all the participants at the meeting, and perhaps more so to the chairmen of the several sessions, that the disagreements did not prevent the delegates from quickly and effectively reaching consensus on the various methods.

There was a clear recognition amongst the delegates that it was essential to produce the best possible tests so that the use of approved pesticides can continue with as little risk as possible to the soil ecosystem. In accepting this, the meeting was also at pains to emphasise that the methods recommended are those which are most appropriate at the moment, bearing in mind the current state of microbiological knowledge. There are, obviously, many other methods, some already published, some yet to come, which might offer advantages. As yet, however, there is a great need to evaluate and refine them and prove their validity as tests to be included in legislative protocols. The delegates at the meeting accepted that the recommended methods have some limitations. This aspect is discussed further in Chapter 15. It was emphasised repeatedly that improvements can only come from intensive research. In the present climate of political pressures to reduce public expenditure such research is under threat. The consequences of any reduction in our efforts to develop protocols to help ensure the safe use of pesticides in agriculture could be severe. It is quite certain that agriculture will continue to depend heavily on pesticides and the new generation of chemicals are more potent than ever before. It goes without saying that this increase in potency brings increased potential for harm to non-target species. Clearly, regulatory schemes must be flexible in order that deletion of obsolete methods and introduction of proven, improved methods with sound scientific bases are easily possible.

If the series of meetings which culminated in Cambridge in 1985 has achieved anything in addition to the published recommendations, it is a clear demonstration that there is a considerable body of committed, experienced scientists who are vitally concerned with the conservation of a safe and healthy environment alongside an efficient, productive agriculture. This concern is sufficiently deep to overcome differences arising from different philo-sophical, geographical and political considerations. The resultant harmony and productivity provides an object lesson for many others and must not be dissipated by insufficient support for the essential appropriate research.

CHAPTER 1

PERSPECTIVES ON SIDE-EFFECTS TESTING

L. Somerville

Pests in the world today are destroying about 35% of all potential food crops before harvest (Pimentel, 1981). These losses are primarily due to insects, plant pathogens and weeds. After the crops are harvested an additional 10-20% are destroyed by insects, microorganisms, rodents and birds. Thus, as much as 48% of the potential world food material is being destroyed annually by pests despite extensive use of pesticides. The argument in support of the continued use of pesticides is, therefore, without question but some constraints may well be required.

While it is difficult to obtain accurate figures for pesticide usage, total world-wide sales of pesticides in 1985 were estimated in terms of end-user market value at $15.9 billion, or in tonnage terms some 3 million tonnes

Table 1   Pesticide sales worldwide - 1986 ($ million)

Area	Herbicides	Insecticides	Fungicides	Others	Total
USA	3100	1090	330	330	4850
W. Europe	1475	850	1100	400	3825
Far East	775	1300	785	90	2950
Lat. America	485	655	250	60	1450
E. Europe	625	450	230	95	1400
Rest	615	655	105	50	1425
World Total	7075	5000	2800	1025	15900

(Wood MacKenzie, 1986). Greatest usage was in North America but Western Europe was only slightly behind mainly as a result of the increased use of fungicides.

Pesticides may be expensive but they are profitable for the farmers and Pimentel et al (1981) has estimated that, in the United States, the average pesticide used on crops returned $4 for every $1 invested.

At present the development of a new pesticide takes some 5-7 years and costs at least £12-15 million excluding the cost of all chemicals which fail to make the grade. This figure does not include provision of a manufacturing plant which could be an additional £5-15 million but, nevertheless, indicates a very substantial investment. Set against a protected patent life of some 16-20 years it is clear that if the product launch does not occur until the 7th year of patent life then the manufacturer has only a relatively short period of time to recover the initial investment, let alone generate profit.

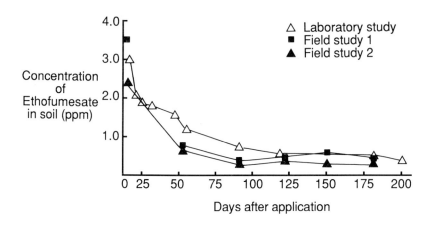

Figure 1   Diagramatic representation of cash flow associated with the development and marketing of a pesticide

As can be seen in Figure 1 investment costs are relatively low in the first years but they increase dramatically with the onset of safety testing, primarily toxicology. The building of manufacturing capacity also requires a substantial capital investment. While some profit is generated in the first 10 years of patent life, it is clear that the manufacturer is looking to the second 10 years to recover his total investment. In this example, the project does not break even until year 15.

One must also consider that, at any time during this 15 year period, resistance to the pesticide can occur or a rival manufacturer can introduce a competitive product. Both situations would seriously impair the profitability of the project.

What does this mean? In simple terms it means that manufacturers take all precautions to reduce the risk associated with their investment. At all stages in the development of the product they are looking at it from a very critical viewpoint. If there are any indications of risk arising from, for example, adverse toxicology, or a toxic impurity generated during the manufacturing process, or long term persistence in the environment, then it may be better to terminate the project at an early stage rather than to continue to invest large sums of money. This critical review is not restricted to the early development stages but extends through to product launch and beyond. The manufacturer jealously guards his reputation. He is never in the market place selling only one product. He has numerous pesticides to offer and he simply cannot afford bad publicity arising from the adverse effects of one product.

Thus, the manufacturers are, in their own way, very demanding regulators of new products both from the point of view of safety to the consumer and safety to the environment. The regulatory authorities only see the small percentage of products which have gained the manufacturer's confidence and do not see the vast number which fall by the way-side.

It has been stated already that large quantities of pesticides are used each year. If the compounds are to be effective then it is essential that adequate coverage of the crop is achieved. The latter normally involves some form of spray application either by tractor, helicopter or fixed wing aircraft although some chemicals are introduced directly into the soil in the form of granules. Various authors have suggested that as little as 1% of the pesticide reaches the target pest. Most of the pesticide, therefore, reaches the non-target sectors of the agricultural ecosystem. The ultimate sink, however, is soil. Clearly there is a need to ensure that the fate of the pesticide in soil is clearly understood, and to this end numerous requirements have been proposed by the regulatory authorities to enable them to evaluate not only the environmental fate of a pesticide but also its potential hazard.

Current regulatory requirements within Europe are outlined in Table 2. In general, requirements are similar for soil degradation and mobility studies and also in the area of aquatic toxicology which are included here as being extremely pertinent to any environmental assessment. Some territories have additional specific requirements such as soil respiration and nitrogen transformation studies which will be discussed further in the course of this volume.

Table 2   Pesticide regulatory requirements in Europe

	UK	France	WG	Holland	Belg.	Den.	Swed.
Soil Degradation	Yes	Yes	Yes	Yes	Yes	Yes	Yes
Soil Leaching	Yes	Yes	Yes	Yes	Yes	Yes	Yes
Soil Adsorption	Yes	Yes	Yes	Yes	Yes	Yes	Yes
Soil Respiration	No	No	Yes	Yes	No	Yes	No
Nitrogen Transformation	No	No	No	Yes	No	Yes	No
Surface Water Degradation	No	No	No	Yes	No	Yes	No
Aquatic Toxicology	Yes	Yes	Yes	Yes	Yes	Yes	Yes

What is meant by degradation studies? These investigations are invariably laboratory studies undertaken to determine the rate of breakdown of the pesticide which, for convenience, is normally in radiolabelled form. The studies are carried out under aerobic, anaerobic and possibly sterile conditions and, generally, more than one soil type is examined. Information obtained would relate to the degree of mineralisation of the pesticide, formation of irreversibly "bound" residues and identification of extractable degradation products. If no breakdown occurs under aerobic, non-sterile conditions then either we are dealing with a 'recalcitrant molecule' or it is toxic to the microbial population. Irrespective of the explanation, further investigation is obviously required. More frequently, however, fairly rapid breakdown and mineralisation is observed and, by comparison with the sterilised soil data, the involvement of microbial activity can be confirmed.

Well designed laboratory degradation studies are an important prerequisite for the generation of good field dissipation data. A comparison of laboratory and field data for the herbicide ethofumesate is given in Figure 2 (Somerville, unpublished data) and, as can be seen good agreement was obtained. Armed with this information the manufacturer gained confidence in laboratory half-life estimates obtained using other soil types and was able to proceed with the introduction of the herbicide into new territories.

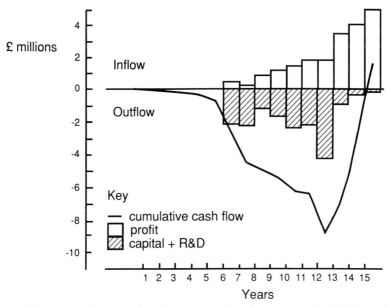

Figure 2   Degradation of ethofumesate in laboratory and field studies

Thus, degradation studies if properly carried out, give meaningful results. The manufacture obtains useful data which enables him to develop his product, and equally important, the regulatory authorities receive good, solid data which they can interprete with confidence. Also, although perhaps of lesser importance, there is a fair degree of uniformity in terms of requirements for degradation studies between authorities, which helps everyone.

Similar comments can be made concerning mobility studies. The test systems are easily used and give reproducible data which are simple to evaluate.

Unfortunately, the assessment of pesticide side-effects is not so easy to carry out. Some regulatory authorities even today have refrained from defining any specific requirements in this area. While The Netherlands accepted the "Recommendations" from the 1979 workshop and introduced a requirement for both soil respiration and nitrogen transformation studies, it was noticeable that at the 1985 Cambridge Workshop, the same authority was calling for a reduction in these requirements. Over the intervening six year period the Dutch experience was that soil respiration data were valid but that in the area of nitrogen transformation the ammonification tests were inadequate. Dr. van Dijk discusses the Dutch viewpoint in detail in Chapter 9. It is noteworthy that although the 1979 recommendations were unanimously endorsed by the various Nations represented at that workshop, only Denmark has followed The Netherlands in introducing similar requirements. Even then it is a fairly recent development.

But why should the assessment of pesticide side-effects be based on soil respiration and nitrogen transformation studies? In a paper reviewing "pesticides effects on non-target soil microorganisms" Anderson, (1978), concluded that these tests are selected because they are easy to do and give measurable results. Perhaps that is so, but are they meaningful?

Soil is a heterogeneous matrix and its microbial activity is related to various factors including climatic conditions and soil type. Most importantly it relates to the populations of micro-organisms (bacteria, fungi, actinomycetes, algae, protozoa and yeasts) present and to the relative proportions of each group. It is reasonable to assume that not only will the microorganisms be having an effect on the pesticide but that the pesticide will be having an effect on the microorganisms. Thus, it is not surprising to find that a fungicide developed for use on cereals has an inhibitory effect on soil fungi. The point at issue is not whether the pesticide inhibits soil microorganisms but whether this is a significant effect. Are there any instances whereby a pesticide has produced an adverse effect which has proved deleterious to soil fertility or crop yield in the long term?

Furthermore, when considering the inhibitory effects of pesticides, the impact of natural phenomena such as freezing and flooding on soil systems must not be forgotten. A brief summary of some environmental factors which influence microbial activity in soil is given in Table 3. Temperature changes can alter not only populations but can change the dominant microbial type.

Similarly drying, flooding or irrigation, can reduce populations by between 50% and 100%. Simple farming techniques, such as ploughing and harrowing, disturb natural populations and it is interesting to note the incorporation of straw also affects populations and can result in decreased crop yields. Liming or heavy treatment with ammonium sulphate fertiliser also produce effects.

Table 3   Environmental factors including influencing microbial activity in soil

Factor	Cause	Effect
Temperature change	Climatic, e.g. freezing thawing.	Fluctuation in functions and populations >50%–99%.
Water potenital	Drying, flooding, irrigation, elevated soil salinity.	Microbial populations reduced by >50%–100% due to low water availability. (Ayanaba et al 1976).
Physical disturbance	Ploughing, harrowing and other mechanical treatments.	Disturb natural micro-habitants, up to 90% depression. (Lynch and Panting, 1980).
pH variations	Liming or heavy treatment of ammonium sulphate fertiliser.	Certain organisms are sensitive to pH change. (Kowalentco et al 1978).
Soil air	Compaction causing reduced channels, pore volume.	Depressions up to 50% if oxygen is not available.
Nutrient supply	Cultural practice, lack of carbon, in the form organic manure or nitrogen in the form of fertiliser.	Depression and stimulation <50%–90%. (Ayanaba et al 1976).

Clearly, depressions of 50% or more, frequently occur under natural conditions. That being the case, what can be achieved by monitoring either soil respiration or nitrogen transformations in the laboratory using the present methodology? Evidence from the literature suggests that the results are somewhat erratic and difficult to interpret.

It should be remembered that the United States Environmental Protection Agency (EPA), in their 1975 draft guidelines, requested information on the effects of pesticides on non-target microorganisms. By the time the final guidelines were introduced in 1982, this requirement had been deleted. The Agency indicated that this change did not result from a lack of interest in microbial effects but merely reflected agreement among scientists both within and outside the Agency that, until useful conclusions can be drawn from properly designed studies, no studies should be required.

Progress has obviously been made since both the 1979 Workshop and the EPA pronouncement in 1982. The new "Recommended Tests for Assessing the Side-effects of Pesticides on the Soil Microflora" (Appendix B) are clearly a step forward, but yet the question of their necessity must still be considered.

There is no doubt that regulatory authorities must monitor pesticides and their possible side-effects. Equally important, test systems are required, but should they be applied in every instance to every product? Perhaps the Regulators could adopt a flexible approach in this area and agree to evaluate each product on a case by case basis. Evaluation cannot be achieved by comparison against a check list but by examining the facts as they relate to that product and its area of use. They should examine the basic data as it relates to the rate and route of breakdown of the chemical in water and soil. Do you find rapid mineralisation or formation of bound residues? Are the degradation products mobile and/or biologically active?

Perhaps they could consider the results of a simple and rapid microtoxicity test such as that proposed by M.P. Greaves (Chapter 13). This test utilises a large number of microorganisms isolated directly from soil. Therefore, it offers the basis of a simple phase I test. If necessary, both the pesticide and its degradation products could be evaluated.

Following an examination of the basic or phase I data by the regulatory authority, and possibly including joint discussions with the manufacturer, an agreed course of action could be established. Thus, in the simplest case, where a pesticide is relatively short-lived being rapidly mineralised, there would be no requirement for further studies involving non-target soil microorganisms.

In a more complex situation where the pesticide or its metabolites is more persistent, there may well be a case for further studies involving for example, soil respiration or nitrification studies or even investigations relating to nitrogen fixation, if the latter were appropriate.

Whatever the answer, let us not forget that the manufacturer as well as the regulatory authority wishes to support pesticides which are environmentally acceptable.

REFERENCES

Anderson, J.R., 1978, in Pesticide Microbiology, edited by I.R. Hill and S.J.L. Wright, (London, Academic Press) p 313.

Ayanaba, A. and Kang, B.T., 1976, Soil Biology and Biochemistry, 8, 313.

Kowalenko, C.G., Ivarson, K.C. and Cameron, D.R., 1978, Soil Biology and Biochemistry, 10, 417.

Lynch, J.M. and Panting, L.M., 1980, Soil Biology and Biochemistry, 12, 29.

Pimentel, D., 1981, in CRC Handbook of Pest Management in Agriculture, Vol I, edited by D. Pimentel, (Boca Raton, Florida, CRC Press) p 3.

Pimentel, D., Krummel, J., Gallahan, D., Hough, J., Merrill, A., Schreiner I., Vittum, P., Koziol, F., Back, E., Yen, D., and Fiance, S., 1981 in CRC Handbook of Pest Management in Agriculture, Vol II, edited by D. Pimentel, (Boca Raton, Florida, CRC Press) p 27.

Wood MacKenzie, 1986, Wood MacKenzie Agrochemical Service, Edinburgh.

CHAPTER 2

NATURAL VARIABILITY IN MICROBIAL ACTIVITIES

K. A. Cook & M. P. Greaves

INTRODUCTION

Microorganisms are of prime importance in the soil environment in the recycling of key elements essential for biological processes and thus for the maintenance of soil fertility. As current agricultural practice involves the application of biologically active molecules to the soil as an integral part of pest control programmes, then it is evident that the effects of such compounds on related non-target organisms should be carefully considered. It is, however, crucial that any such artificially-induced effects are considered in the context of effects induced by natural stresses. If this comparison is not made we run the risk of missing control options which could be of considerable benefit. Such an approach has been suggested by both Greaves et al. (1980) and Domsch et al.(1983) as the only realistic one by which laboratory data may be evaluated. As a result of their survey of relevant data present in the literature Domsch et al. (1983) concluded that depressions of c.90% occurred frequently under natural conditions and that inhibitions for periods up to 60 days were tolerable. However, in general, reported data on natural variation are often incomplete and the methodology is confused. There is, in fact, little consistency in the choice of parameters, the design of field experiments, and the type of statistical analysis used to process results. In both our laboratories we have been concerned with determining the side-effects of pesticides on microbial populations of soils taken from local sites. Hence, in order to be in a position to interpret our data we have determined the natural variability in these soils using a variety of both analy-

15

tical and statistical methods. This paper presents our data and attempts to evaluate the methodology involved in the determination of microbial activity in the field. A more detailed account of these results will be published elsewhere.

METHODS AND MATERIALS

KENT

Field plot design for long term experiments

The Kent plot, on sandy loam, was situated in a grassland field known as Pig Bank near Sittingbourne Research Centre. The site was chosen as it had not been treated with agrochemicals for at least 5 years, had not been grazed, and was not used for access to other areas. In the longer term experiment a 12x12m plot was divided into 9 equal plots of $16m^2$. Each of these plots was divided on paper into 400 small squares and each identified by an x,y coordinate for sampling purposes. On each of the sampling dates coordinates were selected randomly using a computer programme such that a particular coordinate was only sampled once over the period of the experiment. Similarly only 5 of the $16m^2$ plots were sampled (each corner plot and the centre plot). Composite samples were prepared by removing a 10 x 10cm core from each of the 5 areas and mixing them. Five such composite samples were taken. The plot was left uncultivated and uncut throughout the experiment and grass in the sampling area was cut back prior to sampling with a soil auger. All sampling was carried out between 0900 and 1100 hours. A square quadrant subdivided into $0.4m^2$ squares was used to identify the plot coordinates. These experiments were carried out between March 1980 and October 1981.

Field plot design for short term experiments

Two plots, each of $12m^2$ metres were created. One plot was cleared of excess vegetation and the grass cut to approximately 5cm above ground level with a mower. The other plot was dug and rotavated to a depth of 5-10cm. The plots were adjacent to each other and each had a border of 0.5 -0.75m to reduce edge effects. On paper each plot was divided into 300 x $0.4m^2$ squares which were identified by x,y coordinates. During sampling coordinates were identified by a random numbers computer programme and a 10cm core taken to a depth of 5cm with a soil auger. All samples were taken between 0830 and 1000 hours. For sampling purposes the two plots were treated as mirror images of each other. These experiments were carried out between August and September 1984.

Treatment and storage of soil samples

All soil samples were transported to the laboratory in loosely tied black plastic bags. In the laboratory soil samples were spread out in shallow trays and large stones and plant and animal debris removed by hand. Samples were then sieved through a 4mm sieve and, where stated, subsequently through a 2mm sieve.Where samples were too wet to sieve they were allowed to air dry for a minimum period. When in use samples were held at 20°C or at 4°C if overnight storage was required. For longer term storage samples were contained in polyethylene bags loosely secured with rubber bands and stored at 4°C.

Parameters measured

The parameters measured in the two field studies are detailed in Table 1.

Measurement of soil parameters

Field moisture capacity, moisture content, and pH were determined as described by Greaves, et al. (1978).

Determination of combustible carbon

Samples previously dried at 105°C for determination of moisture content were furthr dried at 550°C for 2h. After cooling in a dessicator at room temperature the weight loss was determined. The combustible carbon content was calculated from:

% combustible = loss of weight after heating to 550°C
carbon   --- x100
              weight of initial sample dried at 105°C

Soluble organic carbon content

Soluble organic carbon was determined by the method of Bromley and Cook (unpublished). Soil samples (5g) were placed in sterile plastic universal bottles and 20ml sterile KCl (0.05% w/v) added. Samples were roller mixed for 1h and then centrifuged (6000rpm x 3 min in a Beckman model TJ-6). The supernatant was filter sterilised (cellulose acetate filters 0.2um pore size, paper prefilters – Millipore UK) and the soluble organic carbon content determined using an aqueous carbon analyser (Dohrman DC-80, Sartec Ltd.). Interference was encountered as a result of ions released on extraction of the soil samples. Hence the machine was washed with a volume of 30ml of water between samples and a reagent recommended for use with saline solutions was employed. This comprised $K_2S_2O_8$ (40 g), $HgCl_2$ (16.4g), $Hg(NO_3)_2$, hemihydrate (18.8g) and 10 ml concentrated HNO_3 in 2l distilled water.

Table 1   Parameters measured in the field studies

Oxfordshire	Kent	
	Long term	Short term
Available PO4	Soluble C	Soluble C
Total C (dichromate)	Combustible C	Combustible C
Moisture content	Moisture content	Moisture content
pH	pH	pH
Temperature (soil)	Temperature (air)	Temperature (soil & air)
Rainfall	Rainfall Relative humidity	Rainfall
Urease	Urease	
Ammonia	Ammonia	Ammonia
Nitrate	Nitrate	Nitrate
Nitrite	Nitrite	Nitrite
CO2 evolution	CO2 uptake	CO2 uptake
Dehydrogenase		
Phosphatase		
Fungi	Fungi	
Bacteria (microtitre)	Bacteria (fluores.)	Bacteria (microtitre)
	Yeasts	
	Algae	
	ATP	

The machine was calibrated against a series of standard solutions of sodium hydrogen phthallate and the carbon content of samples determined with a recording integrator. Samples were either analysed immediately or stored at -20°C for up to 4 months. The soluble organic carbon content was calculated from:

$$\mu gC/ \text{g dry soil} = \frac{ppm \times [20 + (\text{soil wet wt} \times [1- \text{dry wt/wet wt}])] \times 1000}{1000 \times \text{dry wt soil used}}$$

Enumeration of microorganisms

1. Differential fluorescence staining

   The method used was that of Anderson and Slinger (1975) using differential fluorescence staining for direct microscopic enumeration of both viable and non-viable microorganisms in soil smears. The stain used comprised a europium chelate, europium (iii) thenoyl trifluoroacetate (Eu[TTA]3) and a fluorescent brightener, the disodium salt of 4,4'bis-(4-anilino-6-bis(2-hydroxyethyl) amino-5-triazin'-2-ylamino)2,2'-stilbene disulphonic acid.

2. Enumeration by plate counting

   Soil (1g) was suspended in sterile distilled water (10ml) and the slurry mixed thoroughly by rolling for 30 min. To remove microorganisms bound to soil particles the slurry was sonicated for 2 min (MSE Soniprep 150 ,2-4 micron amplitude) with periodic cooling. The sonicated slurry was then diluted and viable counts carried out using the spread plate technique. Bacteria, actinomycetes, fungi, yeast and algae were enumerated using the selective media described by Greaves et al. (1978).

3. Enumeration on microtitre plates

   The method was adapted from that described by Darbyshire et al. (1974).

   Soil samples (1g) were suspended in 10ml sterile distilled water in plastic universal bottles (25ml) and roller mixed for 1h. The mixed samples were then sonicated for 5min in a sonicating water bath (Nusonics Ltd.). This was found to be as effective as the use of a conventional sonicator without the associated problems of heat generation. Soil samples were then diluted, if required, prior to distribution on microtitre plates.

Determination of oxygen uptake

Oxygen uptake was measured manometrically as described by Loveridge and Cook (unpublished) using a Gilson differential respirometer.

Soil samples (20g) were weighed into Erlenmeyer flasks (50ml capacity) fitted with centre wells (35 x 8mm). Flasks were preincubated at 20°C for 30 min prior to measurement of oxygen uptake which was then monitored over a 3-4h period. The rate of oxygen uptake was calculated from the initial slope of the uptake curve as below:

$$\mu l \ O_2/g \ dry \ soil/h = \frac{initial \ slope \ of \ curve \ (\mu l/min) \ x \ 60}{dry \ wt \ of \ soil \ used}$$

Estimation of nitrate, nitrite & ammonia

Ammonia, nitrate and nitrite were extracted from soil samples and determined essentially as described by Greaves et al. (1978). KCl (20ml, 3.6% w/v) was added to 10g soil contained in a 25ml universal bottle and the sample mixed by rolling for 1h. Samples were then separated by centrifugation (5000rpm x 5min, Beckman model TJ-6 centrifuge) and the supernatants filter sterilised as described previously. All samples were analysed using an aqueous autoanalyser (Chemlab Insts. Ltd., Essex).

In the longer term experiments carbon analyses were carried out immediately. In the shorter term study daily samples were frozen and all analysed at the end of the study.

ATP determination

Soil samples (3g) were mixed with low response water and agitated for 1h. The resulting slurries were sonicated for 10sec at 8-8.5 micron amplitude using an ultrasonicator (MSE Soniprep 150). The slurry was then diluted 1 to 10 with low response water and ATP extraction and analyses performed essentially as described by Anderson and Davis (1975).

20

Measurement of urease activity

The method used was that of Pettit et al. (1976) but with an incubation temperature of 30°C.

Measurement of soil temperatures

An automatic soil temperature recorder (Grant Insts. Ltd., Cambridge) with 9 probes was used. Probes were inserted to a depth of 10cm at random points over the field plots and the temperature was recorded automatically every hour.

Meteorological data

This was obtained from the monitoring station situated at Sittingbourne Research Centre. Air temperature, relative humidity and rainfall data were recorded daily.

OXFORDSHIRE

The field plot

The Oxfordshire plot was situated at the Weed Research Organisation at Begbroke, Oxford. The site was permanent grass and pesticides had not been used for at least 10 years. The plot measured 10 x 10 m and was rotavated monthly for 9 months prior to use and then at monthly intervals during the growing season for weed control. The soil was a sandy loam type and the soil analysis is shown in Table 2. The experiments described were carried out between April 1977 and May 1978.

Table 2   Analysis of Oxfordshire soil

Coarse sand	26%
Fine sand	40%
Silt	18%
Clay	16%
Total N	0.35%
pH	5.5 - 6.0
Organic C	4%
Available phosphate (ppm)	3.4
Cation exchange capacity	43 (meq/100g)
Field capacity (% water)	27

Nitrogen transformations

Samples were taken to a depth of 15cm from 18 random points (identified using random number tables) within a 400 point grid. Duplicate samples were taken side by side at each point using a 5cm corer. Samples were taken at monthly intervals. Extraction of samples and analysis for nitrate, nitrite and ammonium was carried out as described by Greaves et al. (1978).

Measurement of carbon dioxide evolution

Carbon dioxide evolution was measured in situ at 20 fixed points using 11 Kilner jars inverted over a 20ml sodium hydroxide traps. The Kilner jars were covered with aluminium foil to reduce heating and the traps were changed weekly. Cumulative CO_2 production was measured over a 7 day period by titration of the NaOH as described by Greaves et al. (1978).

Other soil and microbial activity measurements

Additional methods used in the Oxfordshire experiments and detailed in Table 1 have been described by Greaves et al.(1978).

RESULTS AND DISCUSSION

LONG TERM STUDY

Climatic conditions

Climatic conditions over the period of the studies are described in Figures 1 and 2 for Kent and Oxfordshire respectively. A direct comparison of temperature data for the two plots is difficult as air temperature was measured in the Kent study while soil temperature at a depth of 10cm was measured in Oxfordshire. However, the maximum temperatures recorded in Kent in June and November (13°C and 9.4°C respectively) were similar to the soil temperatures in Oxfordshire in the corresponding months (13°C and 5.4°C respectively). Also the maximum temperatures recorded over the experimental period were similar (20.6°C in August in Kent as compared to 19.1°C in July in Oxfordshire). Rainfall was considerably different in the two areas over the experimental period

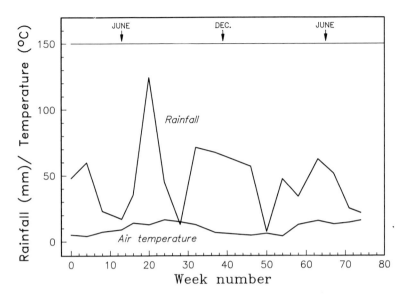

Figure 1   Climatic conditions during the long term study in Kent

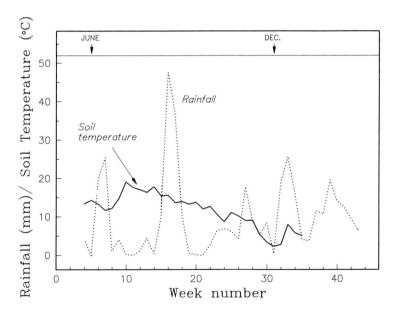

Figure 2   Climatic conditions during the long term study in Oxfordshire

In Oxfordshire rainfall averaged 5-10mm/week over the
entire period with frequent peaks over the winter months
and also notably in the months of June and August.   In
Kent the rainfall was consistently higher with extended
periods of rainfall giving a winter average of about
50mm/month and summer peaks of twice the magnitude
encountered in Oxfordshire.

Respiration and nitrogen transformations

1.   Oxfordshire

Respiration, as measured by carbon dioxide evolution
in the field (Figure 3) was high in the warm wet
conditions prevailing in the summer months, but
declined over the autumn period.   A similar pattern
was observed with the nitrogen transformations,
where nitrate production was at a peak in the summer
months (up to 45ug/g dry soil) and declined in autumn

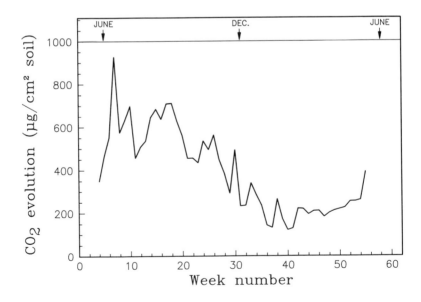

Figure 3   Carbon dioxide evolution - Oxfordshire

(Figure 4). Variation between samples was greatest when the rainfall was heavy, presumably a result of the increased leaching occuring. With both parameters there followed a long recovery period (15-20 weeks) into the next spring before increasing activity levels were observed again. Both activity profiles were characterised by an increase in variability during periods of maximum activity.

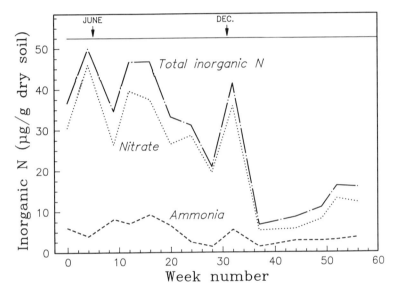

Figure 4   Inorganic nitrogen measurements - Oxfordshire

2.   **Kent**

Little correlation was observed between oxygen uptake (Figure 5) and nitrogen transformations (Figure 6) in the Kent plot. As might be expected oxygen uptake was relatively constant with no obvious seasonal fluctuations. The carbon dioxide evolution data reported above shows that temperature had a large effect on the variability of the data. In the Kent study oxygen uptake was measured at a constant temperature in the laboratory thus masking effects caused by external temperature variations. Similarly the presence of plant cover on the plot could have contributed to the patterns of activity

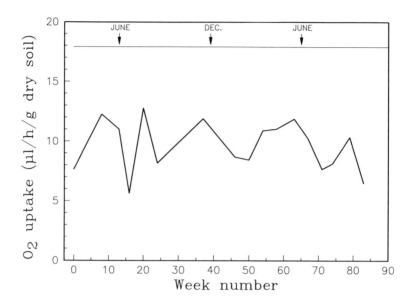

Figure 5   Oxygen uptake measurements - Kent

observed. The nitrogen transformation data was also
difficult to interpret. During the first summer
period levels of inorganic nitrogen were low, but
increased into the second summer period. However,
as plant cover increased over this period the
results could be indicative of the return of plant
nitrogen to the soil. Again the variability of the
data increased as the level of activity increased.
The levels of inorganic nitrogen observed were about
half of those measured on the Oxfordshire plot,
reflecting the previous history of the land used.
Although there was little obvious correlation of
nitrogen levels with respiration measurements there
was a broad correlation of both nitrogen levels and
respiration with soluble organic carbon measurements
(Figure 7). Correlations of soluble carbon with
respiration measurements would, in fact, have been
more pronounced if the May and June readings had
been somewhat higher and it is possible that an
increase could have been missed if it occurred
between sampling periods (ie between weeks 60 and
70).

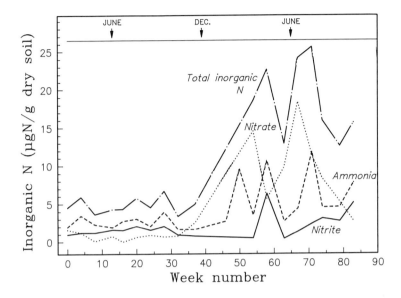

Figure 6   Inorganic nitrogen measurements - Kent

Other parameters

During the course of the longer term studies at both
locations a wide range of parameters were measured (see
Table 1). However, for comparative purposes only the
data on respiration and nitrogen transformations has been
considered in detail. This choice of parameters was
largely on the basis of a complex statistical analysis
(see later) carried out on the Kent data and which
identified respiration and nitrogen transformations as
contributing to a considerable amount of the variation in
the data set. This means that these parameters are more
susceptible to change as a result of external pressures
and are, thus, potentially good criteria to use when
considering influences on microbial activity. It is,
however, of interest to consider briefly the results
obtained with other parameters. The following relates to
results obtained in the Kent experiments. The parameters
measured can be separated into 4 major categories, namely
those concerned with a) climatic variables b) physico-
chemical characteristics of the soil c) microbial
activities and d) microbial numbers.

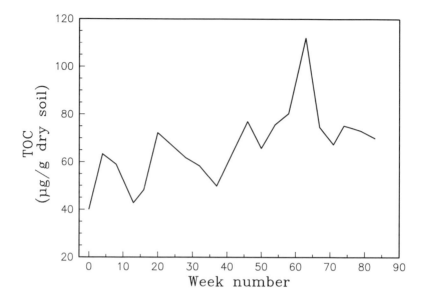

Figure 7   Soluble organic carbon measurements - Kent

a)   Climatic variables such as temperature, rainfall and
      relative humidity, reflect the external environment
      and are, thus, likely to influence conditions in the
      soil itself.

b)   Physico-chemical parameters (moisture content, pH,
      soluble organic carbon, total carbon, phosphate etc)
      describe the soil environment in broad terms and
      are, thus, more likely to correlate with
      microbiological data. However, it should be borne
      in mind that these are still gross characteristics
      which may give little indication of the
      environmental conditions existing in the specific
      micro-environments in which microorganisms work.
      Perhaps one of the best examples of this is pH.

c)   In general, measurements of microbial activity are
      likely to be of prime interest to the agricultural
      microbiologist who is concerned with the
      transformations of chemicals in the soil. In both
      Kent and Oxfordshire plots a variety of soil enzyme
      activities were measured (urease, dehydrogenase,
      phosphatase), but it was found that enzyme levels

28

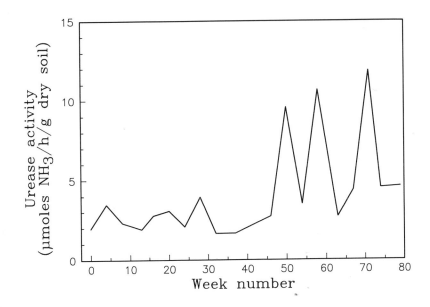

Figure 8   Urease activity measurements - Kent

were not good indicators of microbial activity. Although urease activity (Figure 8) appeared to correlate with both nitrogen transformations and soluble organic carbon measurements (Figure7) to some degree, an analysis of principal components showed that such measurements were actually responsible for very little of the overall variation in the data set. Similar conclusions were drawn with respect to the data obtained by the measurement of other enzyme systems.

d)   Traditional counting methodology is still widely used in ecological studies and data are included here for comparative purposes (Figure 9). No obvious correlations existed between these numerical data and data from other parameters. Furthermore, the Kent study failed to demonstrate any relationship between traditional plate counts and measurements of ATP levels or viable counting by epifluorescence microscopy (Figures 10 and 11),

29

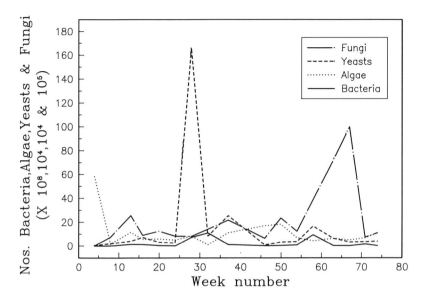

Figure 9 Microbial counts - Kent

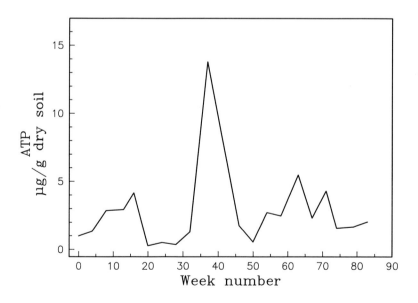

Figure 10   ATP measurements - Kent

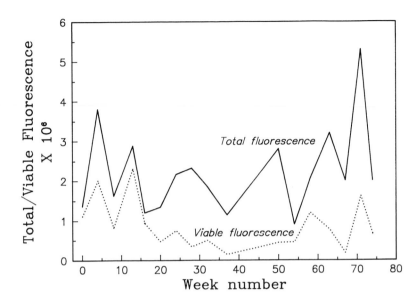

Figure 11   Total and viable counts of microorganisms by fluorescence – Kent

although these parameters were highly correlated in laboratory studies with correlation coefficients in the order of 0.95 (Cook and Potter, unpublished results). Statistical analyses indicated very high variation between replicates and little correlation with any other parameters.

SHORT TERM STUDY

The purpose of the study was a) to consider variation in microbial parameters on a daily basis, b) to look at the effects of plant cover on such parameters and c) to provide sufficient data to allow a more complex statistical analysis to complement that performed as part of the longer term experiments. The parameters measured were chosen on the basis of the previous statistical analyses. Principal components analysis identified those parameters responsible for the bulk of the variation in the data set and hence those that would be expected to be the most sensitive indicators of change.

31

Climatic conditions

A very small amount of rain fell in the first few days of the experiment (1-5mm) followed by a long dry spell where maximum soil temperatures reached as high as 24°C (figure 12). Rainfall increased in September but fell mainly on isolated days causing some decrease in soil temperatures.

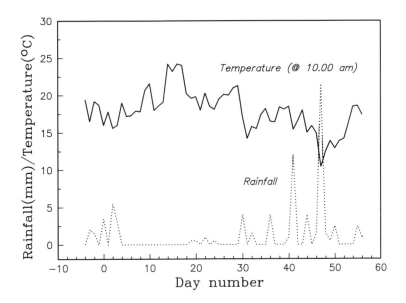

Figure 12   Climatic conditions during the short term study in Kent

Moisture content

This was directly related to rainfall (Figure 13) and was somewhat higher on the grass plot. This could be attributed to a decrease in evaporation from the soil surface caused by the vegetative cover, but one should also consider the expected increase in water uptake due to the vegetation. Equally, increased water loss from soil might be expected as a result of plant trans-piration. Active transpiration can lead to upward movement of water in the soil profile which in turn could compensate for transpirational water loss.

Physico-chemical characteristics of the soil

As expected, combustible carbon, being effectively a measure of structural carbon in the soil, remained approximately constant, at between 13% and 16%. Soil pH was similarly constant, varying between 7.05 and 7.32 over the course of the experiment. Soluble organic carbon levels, however, varied considerably over the trial period (Figure 14) and showed an approximate

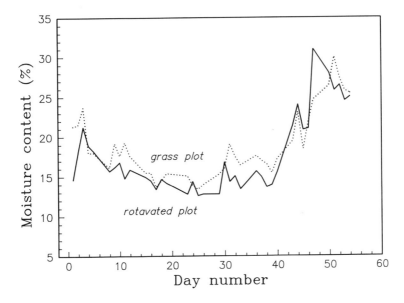

Figure 13   Moisture contents; Kent short term study

inverse relationship to moisture content. A likely explanation is that the soluble carbon pattern reflects the higher metabolic activity in soil at higher moisture contents (see discussion below on oxygen uptake measurements). Soluble carbon levels in the grass plot showed the same trend but with a more pronounced variation. This was probably the result of increased metabolic activity by the vegetative cover and within the rhizosphere as these will be similarly influenced by moisture content.

Microbial activity measurements

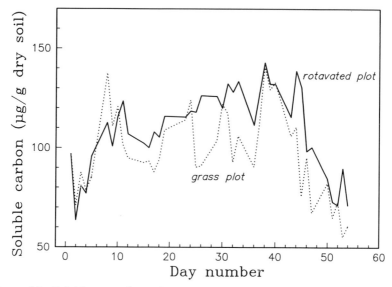

Figure 14 Soluble organic carbon measurements – Kent short term study

Oxygen uptake (Figure 15) appeared to be proportional to moisture content (and thus also to rainfall) and inversely proportional to the soluble organic carbon content of the soil. The higher oxygen uptake infers higher microbial activity resulting in increased carbon metabolism and hence a decrease in soluble carbon present in soil samples. Oxygen uptake on the grass plot was slightly higher than on the rotavated plot. Inorganic nitrogen levels in the two plots were considerably different. On the rotavated plot nitrate comprised the bulk of the inorganic nitrogen present as evidenced by a comparison of nitrate and total inorganic nitrogen levels (Figure 16). However the nitrate levels on the grass plot were considerably lower because of nitrate uptake by the vegetation (Figure 17). Ammonium and nitrite levels were similar for the two plots and showed little fluctuation throughout the experimental period (Figures 16 and 17).

There was a strong indication of a direct relationship between ammonium levels and both oxygen uptake and moisture content. This would be expected as the result of higher overall microbial activity and a correspondingly greater release of ammonium from organic compounds in moist, well-aerated conditions.

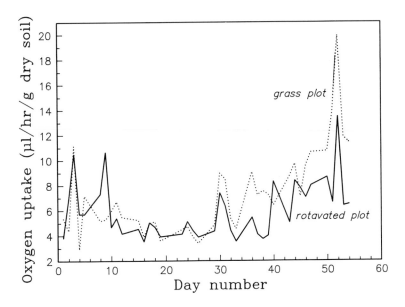

Figure 15   Oxygen uptake measurements – Kent short term study

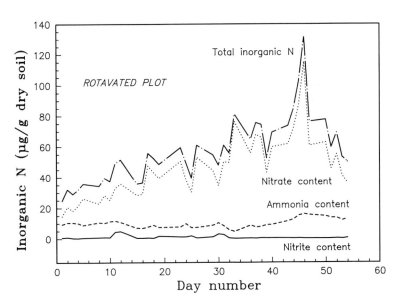

Figure 16   Inorganic nitrogen measurements – rotavated plot,
Kent short term study

In the rotavated plot there was initially an almost linear increase in nitrate, levels falling only towards the end of the experimental period as higher rainfall was encountered. This accumulative increase would be expected in the absence of both plant cover and rainfall, and the final decrease in levels was likely to be attributable to increased leaching. The absence of a similarly distinct pattern on the grass plot was a result of nitrate uptake by the grass cover.

Microbial counts, as measured by the most probable number technique were extremely variable and showed no obvious correlation with any of the other parameters measured (Figure 18). Similar results were obtained in the long term trial and one must conclude that microbial counts are an unreliable parameter to measure in field experiments of the type described.

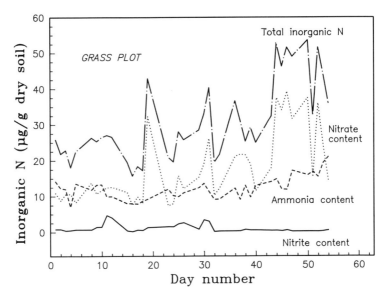

Figure 17   Inorganic nitrogen measurements - grass plot, Kent short term study

Variation of the data

LONG TERM STUDY

Although the data obtained was subjected to a

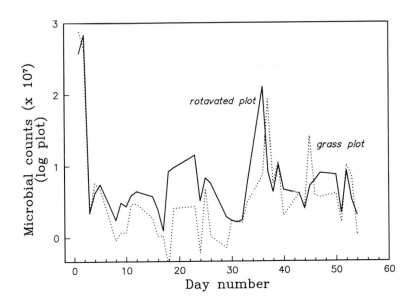

Figure 18   Microbial counts - Kent short term study

detailed statistical analysis (see below) an indication
of the overall variation encountered can be obtained by a
relatively simplistic treatment of the results. In
assessing the variability of the different parameters
measured we are concerned with both spatial and temporal
variation. Spatial variation is the variation between
samples taken at different positions on the plot at a
fixed time and can be expressed as the ratio of the
highest to the lowest values obtained with replicate
samples. This ratio reflects the maximum degree of
variation. The average degree of variation is calculated
as the mean of the ratios obtained for each time point.

Temporal variation can similarly be expressed as the
ratio of the highest and lowest mean values obtained over
the duration of the experiment.

The values reported in Table 3 indicate considerably
higher spatial variation in the Oxfordshire than in the
Kent results, both in comparison of the extreme and the
mean results. This is probably a reflection of the
different procedures used in the two cases. In the Kent

Table 3   Variation in the long term study

| | | OXFORDSHIRE | | | KENT | |
		Extreme	Mean		Extreme	Mean
Variation between replicates	N	7.5	4.0		2.3	1.5
	CO_2	7.9	3.2	O_2	1.8	1.4
Variation over time (between means)	N		7.4			7.6
	CO_2		7.6	O_2		2.3

experiments, pooled soil samples were used in a conscious attempt to remove the spatial variation expected from an uneven sampling site and this procedure is obviously effective in this respect. In addition, the variation encountered in the oxygen uptake results appears to be lower than that for the nitrogen transformation results from the Kent plot. This is probably due to the method used for the measurement of oxygen uptake where experiments were performed in the laboratory under constant conditions. The importance of temperature in influencing respiration is well known.

When we consider temporal variation in the data a number of interesting facts emerge. The temporal variation for nitrogen transformations is remarkably similar at the two separate locations and is also similar to that in the respiration data as measured by carbon dioxide evolution. Variation is reduced considerably when oxygen uptake is considered but, as mentioned above, this parameter was measured under controlled conditions hence removing much of the influence of natural environmental fluctuation.

SHORT TERM STUDY

Spatial variation was somewhat greater than encountered in the longer term experiments, if one considers extreme values, but broadly similar on average (Table 4). While respiratory values were similar, the

Table 4   Variation in the short term study

		ROTAVATED		GRASS	
		Extreme	Mean	Extreme	Mean
Variation between replicates	O$_2$	9.1	2.0	9.7	2.4
	NO$_3$	4.6	2.2	10.9	3.6
Variation over time (between means)	O$_2$	3.9		6.9	
	NO$_3$	7.8		5.3	

variation in nitrate values obtained was considerably greater on the grass plot. This was most likely due to the increased variability resulting from uptake by vegetation on the plot. Nitrate was compared in these experiments as it comprises the bulk of the inorganic nitrogen content of the soil (Figures 16 and 17) and hence closely paralleled total nitrogen levels. Temporal variation on the rotavated plot was generally similar to that seen in the longer term experiments and again the variation in oxygen uptake measurements was low. With the grass plot variation in oxygen uptake was greater reflecting the metabolic activity of roots and organisms associated with the rhizosphere.

Statistical analysis

The results obtained from the studies conducted in Kent were subjected to a detailed statistical analysis which will be fully reported elsewhere. However, certain salient points are summarised below as they relate to the present discussion.

An exploratory data analysis was carried out on the results obtained from the long term experiments largely to determine the interrelationships between the various parameters measured, and the relative importance of the different parameters in contributing to the variation of the data. Principal components analysis was used primarily. Such an analysis can indicate the value of measuring certain variables. Thus, if the bulk of the variation is accounted for in the first few principal components then parameters accounted for in later components may be contributing little to the data variation and hence could be excluded. Similarly, highly correlated variables may be indicating similar trends and may, therefore, be mutually exclusive. Such an analysis can assist in the evaluation of parameters and, hence, in the subsequent design of similar experiments.

The analysis of data from the long term experiments identified nitrogen transformations and respiration as parameters accounting for a considerable variation in the data, while parameters such as urease activity, pH and combustible carbon were of less significance. These conclusions were taken into account in the design of the shorter term experiments and a similar analysis on these results confirmed that the parameters chosen were responsible for the bulk of the variation in the data set. The larger number of data points generated in the short term experiments allowed a more complete analysis to be performed and it is interesting to note that a time shift analysis indicated that a lag of 3–5 days may be appropriate for the response of microbial parameters to changes in the environment.

CONCLUSIONS

The overall objectives of the work reported here were to assess natural fluctuations in the size and activity of microbial populations in the field.Such data are necessary in order to make a reasoned assessment of the real significance of fluctuations observed as a result of chemical treatments in the laboratory. The rational for such an approach has been extensively discussed by Domsch et al. (1983).

From the results obtained it is possible to make a number of observations, both on the overall design and management of field experiments, and on the relationships between environmental conditions and microbial parameters.

If vegetative cover is left on a field plot then any measurements of metabolic activity will include contributions from the vegetation itself and from rhizosphere microbial populations. This obviously makes interpretation of the results more difficult. In addition both physical and chemical characteristics of the soil e.g. moisture content, temperature, and nitrogen status can be affected by plant cover. Hence the determination of specific effects on microorganisms can be complicated. There are also increased problems of plot management associated with the continual growth of the plant cover, and the increased mechanical problems of soil sampling. However, despite these problems, the mean variation observed on the grass plot in these experiments was similar to that seen on the rotavated plot, although the ratios of extreme values were higher.

From the results obtained it is obvious that variation can be decreased by the choice of sampling regimes. Hence, where soil cores were pooled to give composite samples, as practiced in the longer term experiments in Kent, then the spatial variation was much reduced while temporal variation was little affected.

In the course of these studies a large number of parameters were evaluated, but particular emphasis was placed on the value of the parameters measured and their susceptibility to change as a result of environmental influences. In brief, those functions that appeared to be most responsive to environmental change were nitrogen transformations and soil respiration. Of interest here is the use of multivariate analysis in the assessment of variability and this will be the subject of a more detailed publication.

It is apparent from the foregoing discussion that the sensitivity of measurements and the degree of variation encountered in field experimentation can be profoundly affected by the experimental design. It is, thus, possible to choose parameters which are largely unaffected by changes in conditions, and to use a sampling regime which will minimise variation in results. The converse obviously applies. However, in general terms, the most striking feature of the field data presented is the variability encountered even when attempts have been made to minimise this. In the context of side-effect testing this is of considerable importance and emphasises the necessity to adopt an evaluation scheme of the type proposed by Domsch et al.(1983).

It is, in our opinion, crucial that pesticide side-effects are considered against a background of natural variation, and not solely on the results of isolated laboratory tests which may have little basis in reality.

REFERENCES

Anderson, J.R. and Davies, P.I., 1973, in Modern Methods in the Study of Microbial Ecology, edited by T.Rosswall. Bulletin Ecological Research Committee, Natural Science Research Council, Sweden, 17, 271.

Anderson, J.R. and Slinger, J.M., 1975, Soil Biology and Biochemistry, 7, 205.

Darbyshire, J.F., Wheatley, R.E., Greaves, M.P. and Inkson, R.H.E., 1974, Review Ecologie Biologie du Sol, 11, 465.

Domsch, K.H., Jagnow, G. and Anderson, T-H., 1983, Residue Reviews, 86, 66.

Greaves, M.P., Cooper, S.L., Davies, H.A., Marsh, J.A.P., and Wingfield, G.I. 1978. Methods of Analysis for Determining the Effects of Herbicides on Soil Microorganisms and their Activities, Technical Report, Agricultural Research Council, Weed Research Organisation No. 45.

Greaves, M.P., Poole, N.J., Domsch, K.H., Jagnow, G. and Verstraete, W. 1980, Recommended Tests for Assessing the Side-Effects of Pesticides on the Soil Microflora, Technical Report, Agricultural Research Council, Weed Research Organisation No. 59.

Pettit, N.M., Smith, A.R.J., Freedman, R.B. and Burns, R.G., 1976, Soil Biology and Biochemistry, 8, 479.

ACKNOWLEDGMENTS

    The work described here was carried out over a number of years and hence involved the participation of a number of staff from the two organisations involved. We would, however, especially like to thank Jane Potter, Sonja Prinold and Stewart Bromley from Shell Research Ltd. at Sittingbourne, and Sue Greaves, Hazel Davies, John Marsh and George Wingfield from the Weed Research Organisation (now the AFRC Institute of Arable Crops Research at Long Ashton).
    In addition we would like to thank Dr P. North of the Applied Statistics Research Unit at the University of

Kent at Canterbury for his work on multivariate analysis of the data obtained from the field studies in Kent. This work will be reported in more detail elsewhere.

CHAPTER 3

HANDLING AND STORAGE OF SOILS FOR PESTICIDE EXPERIMENTS

J. P. E. Anderson

INTRODUCTION

Study of in situ soil microorganisms in the laboratory

Soils are used in laboratory systems to investigate interactions between microorganisms and plant protection chemicals. Since soils consist of both living and non-living components, and these components invariably occur in different combinations, soils are both complex and heterogenous. In spite of this, laboratory studies with soils can be used to produce environmentally relevant data. There are several prerequisites to production of relevant data: first, the goals of each study must be clearly defined; second, a basic appreciation of the dynamics of the living "component" of the soil, the microflora, is necessary; third, the limitations imposed on the laboratory system by the microflora must be known; finally, the limitations must be observed! This paper deals with the microflora in

FOOTNOTE

This paper was presented as a lecture at the 3rd International Workshop on the "Side Effects of Pesticides on Soil Microorganisms" in Sept./Oct. 1985 in Cambridge. The written version has been held in the form of the lecture, and no attempt has been made to thoroughly review the literature or acknowledge the contributions of colleagues in this and related fields. The data presented here come exclusively from our laboratories and are cited as published or still to be published work.

soils which have been brought for study into the laboratory. Three major aspects will be examined:

1. Effects of soil storage on the microflora and its biochemical functions.
2. Duration of laboratory experiments with soils.
3. Reasons for changes in the microflora in stored soils.

Definition : Microbial Biomass

The standing crop of microflora in soils contains dead and moribund cells, cells that are in a resting state, and cells that are metabolically active and capable of immediate response to changes in the environment. In this paper, the metabolically active portion of the soil microflora is of interest. For convenience, this portion of the microflora will be referred to as the microbial biomass.

EFFECTS OF STORAGE ON THE MICROBIAL BIOMASS AND ITS BIOCHEMICAL FUNCTIONS

Preparation of soils for use in laboratory studies

Representative samples of soil are usually collected from the upper 10 to 20 cm layer of agricultural plots, transported to the laboratory, and adjusted to a water content that allows passage through a sieve. After sieving (e.g. 2 mm) to remove stones, plant residues and larger soil animals, soils are either used immediately in experiments or stored until needed. Can storage influence the microbial biomass and its physiological functions?

Storage of soils at temperatures and moisture conducive to microbial activity and growth

Until quite recently, it was believed that holding "moist" soils at "warm" or "room temperatures" would ensure floristically intact and metabolically active microbial populations. To test the validity of this hypothesis, soils from several countries were collected from the field, transported as rapidly as possible to the laboratory, passed through a 2 mm sieve and stored under aerobic conditions in the dark at 22 $\pm$ 2°C and 40 to 60% water capacity. Periodically, samples of the soils were taken and the amounts of microbial biomass were determined

46

by the physiological method described by Anderson and
Domsch (1978). With this method, rapid and short termed
changes in the microflora can be readily measured
(Anderson et al., 1981).

The results of the experiment (Table 1) show that
storage of soils under conditions conducive to microbial
activity and growth caused substantial losses of
microbial biomass with time. Within 70 days, many soils
had lost up to or more than 50% of their initial weight
of microflora. In further storage experiments at 22 $\pm$
2°C with more than 20 freshly collected, moist, sieved
agricultural soils, including samples from temperate and
tropical zones, it was found that the microbial biomass
similarly decreased with increasing storage time (data
not shown here). Clearly, storage of soils under warm,
moist conditions does not hold the microflora in a steady
state.

Table 1   Loss of microbial biomass in soils stored under aerobic
conditions (in the dark at 22 $\pm$ 2°C and 40 to 60% water
capacity)

Soil	Origin	mg microbial C/kg dry wt soil after:					
		0 Days	7 Days	14 Days	28 Days	70 Days	% decrease (a)
A	Germany	48	36	32	40	34	29
B	Germany	124	132	80	44	30	76
C	Germany	260	250	220	180	70	73
D	England	428	422	404	368	250	42
E	Germany	536	444	340	344	268	50
F	Germany	552	524	396	376	260	53
G	Switzerland	603	508	432	320	240	60
H	Germany	664	592	468	428	390	41
I	Germany	893	949	785	612	600	33
J	Switzerland	933	812	676	592	500	46
K	Germany	981	1085	837	793	700	29

(a) % of initial biomass lost after 70 days storage

Soil moisture and survival of microbial biomass

Perhaps the water content of the soils was not
optimal for microbial survival. To study the influence
of soil water content on biomass vitality, samples of

sieved (2 mm) agricultural soil, a parabrown soil (total C = 1.26%, total N = 0.12%, pH = 5.4) were either, held at their original water content of 12.3%; dampened to 16.4% or 19.0% water content; or dried to 9.0%, 5.5% or 2.4% water content. The soils were stored under aerobic conditions in the dark at 22 ± 2°C and microbial biomass measurements were made with 3 replicate samples immediately after dampening or drying (Day 0), and after 7, 14, 21 and 70 days. Before measuring biomass quantities, the soils were either dried or remoistened to ca. 12% water content and allowed to equilibrate for 4 hours at 22°C.

Table 2 Influence of soil water content on survival of microbial biomass in samples of a parabrown soil (a) stored under aerobic conditions in the dark at 22 ± 2°C. (b)

Length of Storage (Days)	mg microbial C/kg dry wt soil in soil stored at a water content (water potential) of: (c)					
	2.4% (10.6 bar)	5.5% (7.3 bar)	9.0% (5.4 bar)	12.3% (1.8 bar)	16.4% (0.7 bar)	19.0% (0.38 bar)
0	225	350	480	528 (d)	524	528
7	200	375	540	494	440	410
14	200	310	400	388	320	270
21	185	255	330	304	290	250
70	96	125	161	165	145	150

(a) Total carbon =1.26%; total nitrogen = 0.12%; pH = 5.4
(b) After Anderson, 1981
(c) Soils were dried or remoistened to ca. 12% water content and equilibriated for 4 hours at 22°C before making biomass measurements: the original water content of the soil was 12.3%.
(d) Microbial biomass before drying or wetting soil samples = 528 mg microbial C/kg dry wt soil.

The results of the experiment, which are summarised in Table 2, can be used to make two important observations:

1. Regardless of the water content, the microbial biomass in soils held at 22 ± 2°C decreased with

increasing time. Thus, loss of active microflora was not due to a sub-optimal soil water content. (This, incidentally, was also true for the soils listed in Table 1.)

2. Even a short-termed drying of the soil killed measurable quantities of the metabolically active microflora. The initial quantity of microflora in the soils was ca. 528 mg microbial C/kg dry weight soil. On Day 0, after drying samples to 2.4%, 5.5% and 9.0% water content, measurable quantities of microflora were killed. Hence, during preparation of soils for experiments, care should be taken to avoid surface drying of soil aggregates. This kills microorganisms!

Soil temperature and survival of microbial biomass

What are the influences of soil temperature on the survival of the microflora in stored soils? To determine this, moist (ca. 12.4% water content), sieved samples of the parabrown soil were stored under aerobic conditions in the dark at different temperatures. The biomasses in the samples were measured at the start of the experiment, and after 7, 14, 21, 28, 43 and 70 days. Before making measurements, the soils were equilibrated for 4 hours at 22°C.

The results of the experiment (Table 3) show that during storage, the biomass decreased as the temperature increased. Regardless of the temperature, the active microflora eventually decreased with increasing storage time. After 70 days, the least biomass was lost in soils stored at 2 to 4°C; the most was lost at 27°C. These data suggest that if soils must be stored, they should be stored at low temperatures (e.g. 2 to 4°C) regardless of their moisture content.

Biomass and biochemical functions

Storage of soil can induce quantitative changes in the microflora. Population analyses conducted on samples of the parabrown soil stored for 90 days at 20 and 33°C and 12.0 ± 0.5% water content showed that qualitative changes can also occur (Table 4). Under field conditions, in the temperate zone, and especially in agricultural soils subjected to crop rotation, the quantity (Table 5) and doubtlessly also the quality of the soil microflora is in constant flux. Can changes

Table 3   Influence of soil temperature on survival of microbial biomass in samples of a parabrown soil (a) stored under aerobic conditions in the dark at 12.0 + 0.5% water content. (b)

Length of Storage (Days)	mg microbial C/kg dry wt soil in soil stored at a temperature of: (c)					
	2°C	7°C	12°C	17°C	22°C	27°C
0	296	296	296	296	296	296
7	308	292	292	280	288	248
14	292	304	296	288	272	236
28	272	280	256	228	224	168
43	308	268	240	244	184	172
70	244	256	216	188	180	112
% Lost (d)	18	14	27	36	39	62

(a) Total carbon = 1.26%; total nitrogen = 0.12%; pH = 5.4
(b) After Anderson, 1984
(c) Soils equilibriated for 4 hours at 22°C before making biomass measurements.
(d) % of initial biomass lost after 70 days storage

caused by storage influence soils to the extent that they produce ecologically irrelevant metabolic data?

To investigate this question, two sets of data involving a representative bio-chemical function of the microflora, the degradation of a herbicide, will be examined. The first set of data compares dissipation rates of the herbicide in fresh and stored samples of the above described parabrown soil; the second set shows dissipation rates of the herbicide in fresh samples of this soil collected from the field over a 2 year period without regard to season or soil cropping. (Dissipation of the majority of carbamate herbicides from soils is due to microbial degradation (Anderson, 1981, 1984; Anderson and Domsch, 1976, 1980a, 1980b; Banting, 1967; Frehse and Anderson, 1983; Kaufmann, 1967; Smith, 1969, 1970, 1971)).

Herbicide dissipation from stored soils: Untreated samples of the soil were passed through a sieve and stored

Table 4   Fungal species isolated from washed soil particles of fresh
or pre-incubated samples of a parabrown soil. Results are
in per cent of the total populations isolated. (a)

Organisms	Fresh soil	Soil incubated for 90 days: at 20°C	at 33°C
Acremonium furcatum	2.5	c	d
A. roseum	0.3	1.4	1.6
Aspergillus fumigatus	0.3	c	d
Chaetomium difforme	b	1.9	4.0
C. globosum	b	0.5	2.4
Chrysosporium pannorum	b	c	d
Cladorrhinum foecundissimum	3.7	1.9	d
Coniothyrium fuckeli	5.3	c	d
Cylindrocarpon destructans	1.2	c	d
Emericellopsis terricola	0.3	0.5	2.4
Fusarioum avenaceum	0.3	c	d
F. culmorum	1.2	2.4	d
F. equiseti	b	9.1	4.7
F. lateritium	b	4.3	2.4
F. oxysporum	0.9	2.9	6.3
F. sambucinum	2.2	c	d
F. solani	0.3	1.4	1.6
Gliocladium roseum	b	c	1.6
Humicola fuscoatra	b	0.5	4.7
H. lanuginosa	b	c	d
Malbranchea pulchella var. sulfurea	b	c	d
Microdochium bolleyi	1.9	c	d
M. alpina	0.6	1.0	2.4
Mortierella sp. I	7.1	5.3	d
M. stylospora	b	1.9	3.9
M. zonata	1.2	1.4	d
Mucor hiemalis	1.6	c	d
Mycelium sterilum DS VI	3.1	c	d
Penicillium herquei	b	0.5	0.3
P. nigricans	1.2	2.4	3.1
Periconia macrospinosa	b	2.9	0.8
Phialophora cylaminis	1.9	3.9	3.9
Phoma eupyrena	9.6	2.9	d
P. herbarum	0.3	c	4.7
Plectosphaerella cucumerina	b	1.4	d
Pyrenochaeta acicola	b	2.4	0.8
Pythium sp.	4.7	3.8	1.6
Rhinocladiella mansonii	0.6	1.9	d
Trichoderma hamatum	1.2	7.6	6.3
T. harzianum	0.3	2.9	3.1
T. koningii	0.9	2.9	d
Trichosporiella hyalina	2.1	c	d
Verticillium nigrescens	4.3	2.9	2.3

(a) After Anderson 1984

(b) Less than 0.3% of the population

(c) Less than 0.5% of the population

(d) Less than 2.4% of the population

under aerobic conditions in the dark at 20 or 33°C and
12.0 ± 0.5% water content for 90 days. At the end of
this period, the samples were treated with 1 mg a.i. of
herbicide/kg dry wt soil. Freshly collected samples of
the soil were similarly treated with herbicide and all
samples were further incubated for 70 days under aerobic
conditions in the dark at 22 ± 2°C and 12.0 ± 0.5% water
content. The quantities of microbial biomass were
determined, and floristic analyses (Table 4) were
conducted just before addition of the herbicide. At
given intervals after treatment, duplicate samples of the
soil were extracted and the extracts were analysed for
their quantities of parent compound.

Table 6 shows that the rate of dissipation of the
herbicide was highest in the fresh soil. This soil had
an initial biomass of 655 mg microbial C/kg dry wt soil.
Progressively lower rates of herbicide dissipation were

Table 5   Fluctuation of microbial biomass in an agricultural soil
           near Bonn, Germany. (a)

Date of Sampling	mg microbial C per kg dry wt soil	Date of Sampling	mg microbial C per kg dry wt soil
4. VII.83	206	13.VIII.84	330
2.VIII.83	337	10. IX.84	181 (d)
29.VIII.83	165 (b)	15. X.84	319
26. IX.83	320	12. XI.84	284
25. X.83	206	10. XII.84	172 (c)
21. XI.83	132 (c)	5. II.85	273
3. I.84	393	5. III.85	310
13. II.84	331	1. IV.85	426
26. III.84	273	30. IV.85	328
30. IV.84	228	28. V.85	449
21. V.84	280	24. VI.85	411
18. VI.84	286	22. VII.85	458
16. VII.84	210	29.VIII.85	505

(a) Unfertilized soil planted with: Mustard from 13.VI.83 to
     9.VIII.83; Radish from 15.VIII.83 to 10.IX.84; Radish from
     28.V.85 to 17.IX.85.
(b) Drought for more than 14 days
(c) Frozen for more than 14 days
(d) Soil rototilled

found in the soils which had been stored before treatment at 20 and 33°C. At the time of herbicide treatment, these soils had biomasses of 330 and 130 mg microbial C/kg dry wt soil, respectively. Exact analyses of this experiment showed that differences in degradation rates were due to differences in biomass quantities and not to differences in dominant species (Anderson, 1984). Defining DT50 as the time necessary for 50% of the applied dosage of herbicide to dissipate from the soils, it was found that the DT50 value for the thiolcarbamate

Table 6   Persistence of biodegradable thiocarbamate herbicide in freshly collected or stored samples of parabrown soil (a,b)

Days after Herbicide Treatment (d)	Herbicide Recovered in % of Applied (c)		
	Freshly collected soil (e)	Soil stored 90 days before herbicide treatment:	
		at 20°C (e)	at 33°C (e)
0	98.7	96.7	97.9
7	74.4	83.2	87.4
14	59.0	71.1	82.2
28	27.3	59.0	73.9
70	10.0	35.0	58.3
DT50 (f)	ca. 18 days	ca. 42 days	ca. 100 days

(a)   Total carbon = 1.26%; total nitrogen = 0.12%; pH = 5.4.
(b)   After Anderson, 1984
(c)   Herbicide applied at 1 mg a.i./kg dry wt soil
(d)   Soils incubated after herbicide treatment under aerobic conditions in the dark at 12.0 $\pm$ 0.5% water content and 22 $\pm$ 2°C.
(e)   Microbial biomasses in soil (mg microbial C/kg dry wt soil): fresh soil = 655: soil stored at 22°C = 330; soil stored at 33°C = 130.
(f)   DT50 value = Time needed for 50% of the herbicide to dissipate from the soil.

in the fresh soil was ca. 18 days. In the soils at 20 and 33°C, the DT50 values were ca. 42 and ca. 100 days, respectively. Under field conditions, the DT50 values for this particular herbicide range from ca. 14 to 28 days (Fryer and Kirkland, 1970; Kaufman, 1967). Thus, only the value from the fresh soil approached those found in field tests. The results from this experiment indicate that data obtained in microbiological tests with stored soils must be interpreted with caution. Indiscriminate use of values obtained with stored soil to predict field events could lead to serious errors.

Table 7   Persistence of a biodegradable thiolcarbamate herbicide in freshly collected samples of a parabrown soil (a, b)

Date of soil Collection	% of Applied Herbicide Recovered after:			DT50 value (c)	Reference
	7 Days	14 Days	28 Days		
05. IX.1974	74.1	50.4	30.2	14	Anderson (unpubl.)
06. X.1974	76.8	62.5	39.0	22	Anderson (unpubl.)
19. XI.1974	67.6	48.0	37.3	13	Anderson (1981)
13. II.1975	66.6	50.4	30.9	14	Anderson (unpubl.)
08. IV.1975	74.4	-	27.3	18	Anderson/Domsch (1980b)
28. IV.1975	74.4	59.0	27.3	18	Anderson (1984)
11.VII.1975	74.3	63.1	39.8	22	Anderson (unpubl.)
17. IX.1975	70.2	59.4	37.5	19	Anderson (1984)
16. IX.1976	70.5	56.3	36.8	18	Anderson/Domsch (1980a)
Average =	72.1	56.1	34.0	17.5	
Std. deviation	= ± 3.5	± 5.9	± 5.0	± 3.3	

(a) Total carbon = 1.26%; total nitrogen = 0.12%; pH = 5.4
(b) Herbicide applied to all samples at 1 mg a.i./kg dry wt soil: Soils incubated under aerobic conditions in the dark at 12.0 ± 0.5% water content and 22 ± 2°C.
(c) DT50 value = Time needed for 50% of the herbicide to dissipate from the soil.

Herbicide dissipation from fresh samples of soil: Samples of the soil were collected from the field at random intervals between September 1974 and September 1976. Soil was not sampled when it had been frozen or subjected to drought for more than 30 consecutive days. After collection, samples were brought to the laboratory, passed through a 2 mm sieve, and allowed to stand under aerobic conditions in the dark at room temperature (20 to 22°C) in a moist condition (12 to 13% water content) for at least 14 but not more than 21 days. Thereafter, germinated seeds were removed and the soils were treated with the thiolcarbamate herbicide (1 mg a.i./kg) and incubated under aerobic conditions in the dark at 22 $\pm$ 2°C and 12.0 $\pm$ 0.5% water content. At various intervals, samples of the soil were extracted and the extracts were analysed for their quantities of parent compound.

Table 7 shows that regardless of when the parabrown soil was collected, the rates of dissipation were quite consistent. The DT50 values for the herbicide ranged from 13 to 22 days. The average value $\pm$ standard deviation was 17.5 $\pm$ 3.3 days. In contrast to the ca. 42 day and ca. 100 day values obtained with stored samples, the ca. 18 day value with the fresh samples compares favorably to values found in field studies.

Several suggestions can be made from these data:

1. If possible, fresh samples of soil should be used for studying interactions between plant protection chemicals and the soil microflora. This statement is not only valid for investigations on the influence of the microflora on the chemicals, it is equally valid for investigations on the influence of the chemicals on the microflora

2. If soils must be stored, they should be held in a moist condition at e.g. 2 to 4°C. Storage at these temperatures should probably not exceed 3 months.

DURATION OF LABORATORY EXPERIMENTS WITH SOILS

In view of the above data, an important question must be asked: How long can laboratory experiments be conducted with soils before microbiological results become irrelevant? Insight into this question can be obtained from biomass survival tests with unamended and nutrient amended soils.

Unamended soils

Table 1 showed that in 11 unamended soils, the loss of microbial biomass after the relatively short incubation time of 70 days was often greater than 50%. The average loss for the 11 soils was 48 $\pm$ 16%. In tests involving plant protection chemicals and the microflora in sieved, unamended soils, experiments should probably be limited to a maximum of 90 days, or until ca. 50% of the biomass has been lost, or which ever comes first.

Nutrient amended soils

Table 8 shows survival of the microbial biomass in parabrown soil after it had been amended with 0.5% of glucose or 0.5% of a carbohydrate mixture and incubated under aerobic conditions in the dark at 22 $\pm$ 2°C and 12.0 $\pm$ 0.5% water content. In this test, the biomass in the unamended control, as was previously observed (Tables 1, 2

Table 8   Survival of microbial biomass in samples of a parabrown soil
           (a) which received no nutrient supplement or amendment with
           glucose of a plant meal mixture. (b)

Days of Incubation (b)	mg microbial C/kg dry wt. soil in soil amended with:		
	No amend-ment	Glucose (5 g/kg)	Carbohydrate Mixture (d) (5 g/kg)
0	390	390	390
7	380	516	893
14	360	608	930
28	316	527	943
56	160	628	596
70	190	630	524

(a)   C total = 1.26%; N total = 0.12%; pH = 5.4
(b)   After Anderson, 1984
(c)   Soils incubated under aerobic conditions in the dark at 12.0 $\pm$ 0.5% water content and 22 $\pm$ 2°C.
(d)   Carbohydrate mixture: 20% cellulose powder, 40% powdered red clover leaves and stems (Trifolium pratense), 40% powdered wheat straw (Triticum aestivum).

3, and 6, and Anderson et al., 1981), decreased with time. In the nutrient amended soils (Table 8), the biomasses initially grew. After the substrate had been utilized, the biomasses slowly started decreasing. Decrease after growth was particularly evident in the soil amended with the carbohydrate mixture. In tests with amended soils, the length of the test could be determined by survival of the biomass in control samples.

REASONS FOR CHANGES IN THE MICROFLORA IN STORED SOILS

All soils described in this paper were from the temperate zone. In these soils, the average yearly temperatures range from ca. 8 to 15°C. When these soils are brought into the laboratory and held at 22 to 25°C, the activity of the microflora and its rates of metabolism are highly increased (e.g. Table 9). The act of sieving and mixing soil, which is carried through with

Table 9 Influence of incubation temperature on the basal metabolism (unamended soil) of the soil microflora and its mineralization of glucose (4000 mg/kg) in a parabrown soil (a, b)

Incubation Temperature	ml CO_2/100 g dry wt soil/hour (c)	
(°C)	Unamended Soil	Soil plus Glucose
6	0.5	3.8
12	1.2	7.0
18	2.4	12.0
22	3.0	15.8
26	4.0	20.4
30	5.3	28.6
36	5.5	29.0
43	8.6	22.0

(a)   Total carbon = 1.25%; total nitrogen = 0.12%; pH = 5.4.
(b)   Soil samples were incubated at the given temperatures for 4 hours before making respiration measurements: glucose was mixed into the precooled or prewarmed soil and measurements were made during the second hour after amendment.
(c)   Respiration data are averages from 4 replicate 100 g samples.

the good intention of making soil less heterogenous, intentionally removes partially-degraded plant litter and other sources of readily degradable carbon. The main sources of carbon for the soil microflora are then the highly polymerized carbon compounds of the soil, and the freshly dead soil microorganisms themselves. Since incubation at high temperatures greatly increases microbial requirements for available carbon (for maintenance metabolism), but does not increase the availability of the highly polymerized soil carbon (much of which can have an average age of 1000 years!) we are essentially working with carbon limited systems. The primary reason for biomass loss in sieved soils is carbon starvation!

This contention is supported by the data in Table 8, which show that addition of nutrients to the sieved soil counteracts the loss of microbial biomass, and the data in Table 9, which show how drastically rates of carbon metabolism in unamended and glucose amended soils increase with increasing temperatures. It must be stressed at this point, that the addition of nutrients by no means guarantees that shifts in the composition of microflora do not occur. On the contrary, it probably ensures that shifts do occur.

Currently available data indicate that the progressive losses of biomass seen in unsupplemented, stored soils in the laboratory do not commonly occur in planted field soils. Without a doubt, fluctuations in biomass quantities can be measured (Table 5), but there are increases as well as decreases. Under field conditions, plant exudates and plant parts make major contributions to the maintenance of the soil microflora.

SUMMARY AND RECOMMENDATIONS

Laboratory tests can be effectively used to investigate interactions between soil microorganisms and plant protection chemicals. To do this, the limitations of each laboratory system or "model" must be known and observed. Tests involving plant protection chemicals and soil microorganisms involve two equally important aspects: the influence of the chemicals on the microflora and its biochemical processes; and the influences of the biochemical processes of the microflora on the chemical. To ensure production of environmentally relevant data, soils used in tests should contain a microflora which is comparable to that found in the field.

This paper has shown that storage of soils under

warm, moist conditions can rapidly induce changes in the quantity and quality of the microflora, and in the quality of at least one important biochemical process, the microbial degradation of herbicides. The changes in the microflora appear to be due to depletion of readily available carbon. Addition of nutrients to stored soils can counteract losses of biomass, but shifts in microbial dominance patterns can probably not be avoided. Changes in the quality of general biochemical reactions, as shown in the example presented here, have more to do with changes in rate than with changes in metabolite production (Anderson, 1984).

In view of the above discussion, the following are recommended:

1. For routine laboratory tests with the soil microflora and plant protection chemicals, freshly collected soils should be used where possible.

2. During preparation of fresh soils for tests, procedures which damage the microflora (i.e., excessive drying) should be avoided.

3. Before sampling soils, climatic conditions should be considered: soils subjected for longer periods (e.g. 30 consecutive days) to freezing or drying should be avoided.

4. If storage of sieved soils is unavoidable, temperatures of 2 to 4°C for up to 3 months can be used.

5. Routine laboratory experiments with unamended soils (i.e. no nutrients added) should not run for more than 3 months, or until 50% of the initial biomass has been lost, or whichever comes first.

REFERENCES

Anderson, J.P.E., 1981, Soil Biology and Biochemistry, 13, 155.

Anderson, J.P.E., 1984, Soil Biology and Biochemistry, 16, 483.

Anderson, J.P.E., Armstrong, R.A. and Smith, S.N., 1981, Soil Biology and Biochemistry, 13, 153.

Anderson, J.P.E. and Domsch, K.H., 1976, Archives of Environmental Contamination and Toxicology, 4, 1.

Anderson, J.P.E. and Domsch, K.H., 1978, Soil Biology and Biochemistry, 10, 215.

Anderson, J.P.E. and Domsch, K.H., 1980a, Archives of Environmental Contamination and Toxicology, 9, 115.

Anderson, J.P.E. and Domsch, K.H., 1980b, Archives of Environmental Contamination and Toxicology, 9, 259.

Banting, J.D., 1967, Weed Research, 7, 302.

Frehse, H. and Anderson, J.P.E., 1983, in IUPAC Pesticide Chemistry, Human Welfare and the Environment, Pesticide Residues and Formulation Chemistry, edited by Greenhalg, R. and Drescher, N., 4, p. 23.

Fryer, J.D. and Kirkland, K., 1970, Weed Research, 10, 133.

Kaufman, D.D., 1967, Journal of Agricultural and Food Chemistry, 15, 582.

Smith, A.E., 1969, Weed Research, 9, 306.

Smith, A.E., 1970, Weed Research, 10, 331.

Smith, A.E., 1971, Weed Science, 19, 536.

CHAPTER 4

LABORATORY MEASUREMENTS ON SOIL RESPIRATION

J. W. Vonk & D. Barug

INTRODUCTION

The present paper deals with methodology and experience on the effects of pesticides on soil respiration. A definition of soil respiration has been given by Anderson (1982). He defines this process as the uptake of oxygen or the release of CO_2 by living metabolizing entities in the soil. Generally the term soil respiration is used for the microbial contribution to soil respiration, that is by bacterial, fungal, algal and protozoan cells.

In the following we will restrict ourselves mainly to soil respiration studies under laboratory conditions aimed at assessing potential effects of chemicals on soil microorganisms. Soil respiration may be measured either by the uptake of oxygen or by the evolution of CO_2.

OXYGEN CONSUMPTION

Oxygen consumption can be determined directly by measuring the decrease in oxygen gas concentration in a closed soil/air system. This can be accomplished by oxygen gas electrodes or by gas chromatography.

More frequently used are respirometers which can be placed into two categories: those in which the partial pressure of oxygen in the gas mixture continuously decreases and those in which the oxygen partial pressure is maintained at a more constant level. In all respirometers the CO_2 evolved is removed from the gas atmosphere and the net result is a decrease in partial pressure of oxygen and hence a decrease in the gas volume.

A Warburg type of instrument can be modified to measure soil oxygen consumption. The pressure in a flask provided with a soil sample is kept constant by adjusting the water level in a vertical glass tube connected flexibly to the flask. Corrections should be made for atmospheric pressure changes with a thermobarometer (Klein et al., 1972).

In the oxygen-consumption respirometer the gas taken up by the soil is volumetrically replaced. If oxygen is consumed, the pressure in the closed system drops and this activates a contact switch, which controls an electrolysis cell in which the consumed oxygen is replaced (Anderson, 1982).

One of the main disadvantages of the respirometer technique is that no CO_2 is present in the atmosphere, which causes a very low concentration in the soil's gas phase. This may have an effect on the metabolic activity of soil microorganisms.

In all methods where consumed oxygen is not replaced, care should be taken to avoid that the oxygen pressure becomes too low.

CARBON DIOXIDE RELEASE

Measurement of CO_2 released from soils may be accomplished by different methods. CO_2 may be trapped in potassium hydroxide and determined by gravimetric, titrimetric or conductometric methods.

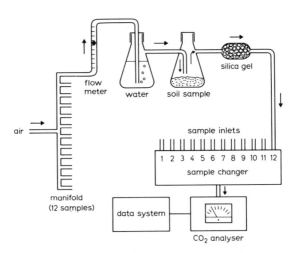

Figure 1   Scheme of the equipment for measuring carbon dioxide evolution from soil

Direct measurement of the CO_2 evolving from the soil can be carried out with gas chromatography. For accurate measurements only closed bottles with a sampling port for syringes can be used, because a flow-through system would reduce the CO_2 concentration too much for accurate measurement. Determination of CO_2 concentrations in gas flowing out of the soil sample flask can be accomplished by infra-red analysis. This system has been used in our laboratory to measure effects of pesticides on microbial soil respiration.

The equipment developed at our Institute is depicted in Figure 1. Compressed air is first saturated with water and than led over batches of 100 g of the soil, held at 0.32 bar moisture pressure. Thereafter the air stream is passed through silica gel to prevent condensation in the measuring system. The CO_2 concentration is measured in an infra-red gas analyser. Every 2 hours the concentration of CO_2 in the gas stream was measured.

Table 1   Concentrations   of   herbicides   with
stimulatory effects on soil respiration

Herbicide	Conc. of herbicides (mg/kg) with stimulation			
	Basal	Lucerne meal amended	Glucose amended	
			Sand	Loam
A	25(a)	25(b)	>2.5	25
B	N.E.	N.E.	25	N.E.
C	N.E.	N.E.	25	N.E.
D	N.E.	N.E.	N.D.	N.D.
E	N.E.	N.E.	N.D.	N.D.
F	N.E.	N.E.	N.D.	N.D.

N.D. = not determined   (a) = 10 days stimulation
N.E. = no significant effect (b) = loam

EFFECT OF PESTICIDES ON SOIL RESPIRATION

Evolution of CO_2 from soil amended with concentrations of pesticides equivalent to field rate and ten times the field rate, is compared with that from a control soil. We used two soils of a different nature: a slightly acid humic sand and a neutral loam soil.

Greaves et al. (1980) recommended to investigate the effect of pesticides on basal soil respiration, that is: without any additional C-source amendment, and on respiration stimulated with lucerne meal. In addition, the Dutch authorities required also tests with respiration stimulated by glucose and ammonium sulphate.

Some of our results are reported here. Since most experiments were carried out as contract research, names of pesticides cannot be given in the following results.

Herbicides

Table 1 lists effects of herbicides tested on soil respiration. The first two columns give the lowest concentration of the compounds with an effect, recorded during long-term respiration experiments. Basal respiration and lucerne meal-stimulated respiration were measured. The last two columns show effects in short-term respiration experiments, in which glucose and ammonium sulphate were added.

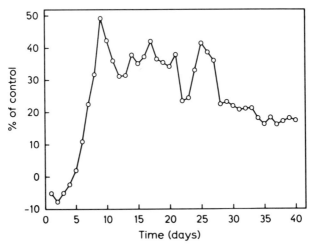

Figure 2   Effect of herbicide A (25 mg/kg soil) on long-term soil respiration of loam soil amended with lucerne meal (0.5%)

Basal activity was only affected by one of the herbicides. It stimulated respiration. This effect was even more pronounced when lucerne meal was added to the soil. On the other hand, all the compounds tested had an effect on short-term respiration. All these effects were a stimulation of respiration. Compound A is known to have a respiration stimulatory effect on microorganisms due to uncoupling of oxidative phosphorylation.

In Figure 2 the effect of compound A on soil respiration during a 40-days incubation period in lucerne meal-amended soil is given as percentage of the control soil. It appears that the stimulating effect is still present after 40 days. An effect above 15% is generally considered to be significant.

Fungicides and insecticides

Table 2 lists the effects of 5 fungicides and 1 insecticide (compound L). In this case two compounds (I and J) gave effects. One of the compounds exerted an inhibition of soil respiration over a period of 20 days if lucerne meal was added to the soil. The other stimulated respiration. The inhibitory effect of compound J on soil respiration is given in Figure 3.

Table 2   Concentrations of fungicides and insecticide (L) with effects on soil respiration

Compound	Concentration of compound (mg/kg) with effect	
	Basal respiration	Lucerne meal amended
G	N.E.	N.E.
H	N.E.	N.E.
I	25 (stimulation)(a)	N.E.
J	N.E.	25 (inhibition)(b)
K	N.E.	N.E.
L	N.E.	N.E.

N.E. = no significant effect
(a)   = 13 days stimulation in sandy soil
(b)   = Sandy, soil

No studies of glucose mineralisation were carried out in this case because such studies were no longer required for registration at that time.

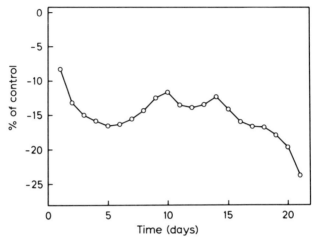

Figure 3   Effect of fungicide J (25 mg/kg soil) on long-term soil respiration of humic sandy soil amended with lucerne meal (0.5%)

Short-term soil respiration

Nevertheless we had found that short-term soil respiration was affected by a number of pesticides (Table 1). Therefore we wanted to investigate whether the simple Warburg respirometer technique was usable to detect effects of chemicals on soil microbial populations.

Table 3   Lowest concentrations (mg/kg) showing significant reductions of respiration in Warburg respirometer tests

Chemical	Sand	Loam
Pentachlorophenol	>1000	1000
2,4,5-T	1000	>1000
Carbendazim	>1000	>1000
Toluene	1000	1000
Tetrachloroethylene	2000	2000

We measured the oxygen consumption by basal respiration, and glucose-stimulated respiration. We investigated the influence of 3 pesticides, namely pentachlorophenol (a general biocide), carbendazim (a fungicide) and 2,4,5-trichlorophenoxyacetic acid (a herbicide), as well as two environmental pollutants: toluene and tetrachloroethylene.

Table 3 indicates the concentrations of these chemicals, at which significant effects on soil respiration could be observed. From the table it appears that these concentrations are rather high, even in the case of pentachlorophenol, a compound known to have a general biocidal effect. However, such concentrations are not unrealistics if these chemicals are spilled on the soil.

COMPARISON WITH LITERATURE DATA

The effects of pesticides on non-target soil microorganisms have been reviewed by Anderson (1978). As far as herbicides are concerned, he noted that cases of inhibition of soil respiration were reported approximately as often as cases where no effect or simulation was observed. For fungicides, effects on soil respiration were generally inhibitory. For insecticides, more stimulation than inhibition was observed.

Recently, Malkomes (1985) reported the effect of sodium trichloroacetate in doses of 43 and 430 mg per kg in two soil respiration tests with and without lucerne meal. Only significant effects were observed for the lucerne stimulated respiration in one soil type at the 10-fold dose (430 mg/kg).

Helweg (1986) treated a coarse sand soil in the laboratory with two herbicides, three fungicides and one insecticide and measured basal and lucerne meal-stimulated soil respiration. This combination of pesticides had no effect on basal respiration, but the lucerne meal-stimulated respiration was inhibited for about 25% in comparison with the untreated soil, for a period of 30 days.

CONCLUSIONS

From a number of arbitrarily chosen pesticides only a few had effects on soil respiration, mostly at levels ten times higher than the normal field dose rate. It is remarkable that generally more effects are found with glucose or lucerne amended respiration than with basal respiration.

Although soil respiration is obviously not a very sensitive process for assessing potential side effects on soil microflora, in our opinion tests with pesticides on soil respiration are useful to screen for compounds that really could destroy our soil microbial community.

REFERENCES

Anderson, J.P.E., 1982, in Methods of Soil Analysis, Part 2, edited by A.L. Page, R.H. Miller and D.R. Keeney (Madison, American Society of Agronomy) p. 831.

Anderson, J.R., 1978, in Pesticide Microbiology, edited by I.R. Hill and S.J.L. Wright (London, Academic Press) p. 313.

Greaves, M.P., Poole, N.J., Domsch, K.H., Jagnow, G. and Verstraete, W., 1980, Recommended Tests for Assessing the Side-Effects of Pesticides on the Soil Microflora. Technical Report Agricultural Research Council, Weed Research Organisation, No. 59.

Helweg, A., 1986, in Microbial Communities in Soil, edited by V. Jensen, and H. Sorensen (Copenhagen, FEMS).

Klein, D.A., Mayeux, P.A. and Seaman, S.L., 1972, Plant and Soil 36, 177.

Malkomes, H.P., 1985, in Behaviour and Side Effects of Pesticides in Soil, edited by M. Hascoet, H. Schuepp, and E. Steen (Paris, INRA) p. 191.

CHAPTER 5

CARBON MINERALISATION: RING TEST RESULTS

J. P. E. Anderson

INTRODUCTION

Carbon cycle

Carbon is the most important single element in the biological realm. It is the building block of all forms of life and the chemical bonds it forms with itself and other elements provide the basis for biochemical energetics. Cycling of carbon in the biological realm can be said to "start" with the fixation of carbon dioxide by green plants and "end" with the release of carbon dioxide during the metabolism of animals, plants and microorganisms. The cycling of carbon is essential to all life.

Carbon mineralization as a test parameter

In soil, one of the most important functions of the microflora is the degradation and partial mineralization of organic matter. This process releases carbon dioxide to the atmosphere and provides energy and carbon to the microflora. In addition, it releases mineral elements which are essential to plant and microbial growth, and leaves stabilized carbonaceous residues which form the basis of the organic matter of soil. The essential nature of carbon mineralization in soils has made this process ideal for use in tests designed to determine the possible influences of pesticides on soil quality.

Definition: Soil Respiration

Carbon mineralization in soil can be readily followed

by measurement of the uptake of oxygen or the release of carbon dioxide. These energy yielding processes, when conducted by living, metabolizing entities in the soil, are termed "soil respiration".

Recommended test procedures

Participants of earlier Workshops on the "Side Effects of Pesticides on the Soil Microflora" recognized the elegance and general nature of experiments on soil respiration and recommended several tests as a standard procedures. Since their publication in 1980 (Greaves et al., 1980) a great deal of experience has been gained using these tests. This paper summarizes the results from a ring experiment with these tests which was conducted by a group of 5 industrial research laboratories. Of interest in this experiment was the practicability of the tests, and the comparability and reproducibility of results.

RING EXPERIMENT TO EVALUATE TEST PROCEDURES

Participants

The companies and research workers that participated in the ring experiment, listed alphabetically, were BASF AG (R. Hamm), Bayer AG (J.P.E. Anderson and G. Hermann), Celamerck (D. Eichler), Ciba-Geigy (J.A. Guth) and Hoechst AG (N. Taubel).

Description of methods used to measure soil respiration

One research group used an automatic device to measure the oxygen consumption of soil samples. Other groups trapped carbon dioxide released from soils with solid (soda lime) or aqueous (NaOH) alkali. The quantities of carbon dioxide trapped by the alkali were determined by weighing (soda lime), titration (aqueous alkali solutions) or colorimetry (aqueous alkali, Autoanalyzer, a Technicon Procedure). All data were expressed as mg carbon dioxide/100 g dry wt soil. Those interested in detailed discussions of methods and equipment are referred to recent review articles (e.g. Anderson, 1982; Greaves et al., 1978).

Design of ring experiments

The soil used in the ring experiment, which was

loamy sand from Field F of the Laacherhof Research Station (Bayer AG), had the following characteristics: total carbon = 0.84%; total nitrogen = 0.08%; pH = 5.3. The soil was collected from the upper 10 cm of the field, brought to the laboratory, passed through a 2 mm sieve, packed in perforated plastic bags, and immediately shipped to the participants. Within a predetermined time span, experiments were started by all participants. In other tests, a sandy silt soil (total carbon = 2.6%; total nitrogen = 0.24%; pH = 5.4) was also used.

Table 1   Ring experiment conducted by 5 laboratories showing the mineralization of lucerne-grass-green-meal (0.5%) in fresh samples of a loamy sand soil (a)

Days of Incubation (b)	Carbon Mineralized: Cumulative Values (mg Carbon Dioxide/100 g dry wt soil) Laboratory: (c)					Average ± Standard Deviation	% Deviation from the Average (d)
	A	B	C	D	E		
1	82	71	48	59	54	63 ± 14	22
2	147	259	83	207	83	116 ± 35	30
4	242	196	206	174	147	193 ± 36	18
7	306	257	261	209	207	248 ± 41	17
14	403	339	344	302	259	329 ± 53	16
21	448	387	399	338	290	372 ± 60	16
28	481	420	406	388	309	401 ± 62	15

Average of (a) = 19%

(a) Total carbon = 0.84%; total nitrogen = 0.08%; pH = 5.3
(b) Soils incubated under aerobic conditions in the dark at 40 ± 5% water capacity and 20 ± 2°C.
(c) The letters were randomly assigned to the laboratories.
(d) % Deviation from the Average = Standard Deviation/Average x 100.

The reference chemical in the tests was mercuric chloride. In the ring experiment, this material was mixed into the soil at rates of 100 and 200 mg/kg dry wt soil.

The nutrient source used in the tests was commercially available lucerne-grass-green-meal. Samples from one charge of this material were powdered and distributed to all participants in the Ring Experiment.

Selected results with nutrient amended soils

Variation from laboratory to laboratory: The goal of one experiment was to compare the respiration rates obtained by each laboratory with nutrient-amended but otherwise untreated samples of the loamy sand soil. Samples of the soil were treated with 0.5% lucerne-grass-green- meal and incubated under aerobic conditions in the dark at $40 \pm 5\%$ water capacity and $20 \pm 2°C$. Respiration measurements were made after 1, 2, 4, 7, 14, 21 and 28 days of incubation.

The results of the experiment, which are given as cumulative values in Table 1, show the high degree of variation from laboratory to laboratory. Deviations from the average respiration values were highest at the start of the experiments but decreased as the experiment proceeded. For Days 1 and 2 of the test, the deviations from the average values were 22 and 30%, respectively. Thereafter, the deviations fell stepwise from 18 to 15%. The overall average deviation for the entire experiment was $\pm 19\%$. This deviation, which would seen quite large, is not uncommon for ring experiments dealing with biological materials.

Inhibition of soil respiration by the reference chemical: The goal of this experiment was to examine the comparability of data from different laboratories in tests in which the reference chemical was used to partially inhibit the mineralization of lucerne-grass-green- meal. For this purpose, soil samples were amended with the plant meal (0.5%) and treated with 100 or 200 mg reference chemical/kg dry wt soil. Incubation was under aerobic conditions in the dark at $40 \pm 5\%$ water capacity and $20 \pm 2°C$. The amounts of oxygen consumed by the soil or the quantities of carbon dioxide released were determined after 1, 2, 4, 7, 14, 21 and 28 days. Since the variation in raw data (i.e. mg carbon dioxide/100 g dry wt soil/unit time) from the laboratories was so large that direct comparisons were impractical, the % inhibition of mineralization was calculated: The calculation was: % Inhibition of Mineralization = 100 −(Treated/Untreated) x 100

Table 2   Ring experiment conducted by 5 laboratories showing the % inhibition of mineralization of lucerne-grass-green-meal (0.5%) in fresh samples of a loamy sand soil (a)

Days of Incubation (b)	Inhibition of Mineralization in % of the Untreated Control (c) Laboratory: (d)					Average ± Standard Deviation	% Deviation from the Average (e)
	A	B	C	D	E		
1	96	93	77	92	98	91 ± 8	8
2	95	96	42	92	100	83 ± 24	29
4	82	62	38	76	98	71 ± 23	23
7	63	42	21	57	80	53 ± 22	42
14	55	28	15	40	58	39 ± 18	46
21	47	28	7	28	46	31 ± 16	52
28	45	29	7	19	38	28 ± 15	53

(a) Total carbon = 0.84%; total nitrogen = 0.08%; pH = 5.3
(b) Soils incubated under aerobic conditions in the dark at 40 ± 5% water capacity and 20 ± 2 °C.
(c) Calculated from cumulative values.
(d) The letters were randomly assigned to the laboratories.
(e) % Deviation from the Average = Standard Deviation/Average x 100.

The results of the experiments are summarized in Tables 2 and 3. Comparison of the data in the two tables shows that there was a concentration dependent inhibition of soil respiration. This is not surprising since the toxic action of mercuric chloride is not selective. Recoveries of soil respiration started almost immediately after treatment and continued throughout the experiments.

Table 3　　Ring experiment conducted by 5 laboratories showing the %
inhibition of mineralization of lucerne-grass-green-meal
(0.5%) in fresh samples of a loamy sand soil (a) after
treatment with 200 mg mercuric chloride/kg dry wt soil

Days of Incubation (b)	Inhibition of Mineralization in % of the Untreated Control (c) Laboratory: (d)					Average ± Standard Deviation	% Deviation from the Average (e)
	A	B	C	D	E		
1	88	94	85	93	98	92 + 5	5
2	89	95	83	92	99	92 + 6	7
4	90	92	77	87	97	89 + 7	8
7	89	84	61	81	95	82 + 13	16
14	88	63	34	50	91	65 + 24	37
21	84	51	29	21	84	54 + 30	56
28	79	46	29	9	73	47 + 29	62

(a) Total carbon = 0.84%; total nitrogen = 0.08%; pH = 5.3
(b) Soils incubated under aerobic conditions in the dark at 40 ± 5% water
capacity and 20 ± 2°C.
(c) Calculated from cumulative values.
(d) The letters were randomly assigned to the laboratories.
(e) % Deviation from the Average = Standard Deviation/Average x 100.

　　　　Comparison of the data from 5 different laboratories
shows that at each time interval, and for either dosage
of reference chemical, there were large deviations from
the average values.　In both experiments, the degree of
deviation from the average increased as the experiments
progressed.　These data show that comparisons of data
from different laboratories, even using a single soil,
are not always feasible.　Of special relevance is that
the data obtained using a reference chemical, even when
expressed as % inhibition in reference to the control,
could not be compared.

Reproducibility of results: In view of the high degree of variation shown when different laboratories conduct a single type of respiration experiment, it is logical to ask if repetitions in a single laboratory give comparable data. To answer this question, data from

Table 4   Experiments conducted in a single laboratory showing the mineralization of lucerne-grass-green-meal (0.5%) in samples of a loamy sand soil (a) collected over a 4 year period without regard to season or soil cropping. (b)

Date of Soil Collection	Carbon Mineralization: Cumulative Values (mg Carbon Dioxide/100 g dry wt Soil)			
	7	14	21	28
15 VI 81	227	278	296	306
4 IX 81	236	285	308	322
1 VI 82	277	313	–	346
24 IX 82	248	293	317	337
28 IX 82	207	259	290	309
16 XI 82	208	281	329	358
29 III 83	215	288	332	363
7 IV 83	217	289	329	359
29 IX 83	215	276	314	342
29 XI 83	243	310	348	378
2 II 84	207	280	320	351
18 III 85	195	259	302	331
24 V 85	249	320	354	390
31 VII 85	257	321	360	391
Average ± Std Deviation	227 ± 23	289 ± 20	332 ± 21	349 ± 26
% Deviation from the Average (c)	± 10%	± 7%	± 7%	± 7%

(a) Total carbon = 0.84%; total nitrogen = 0.08%; pH = 5.3
(b) Soils incubated under aerobic conditions in the dark at 40 ± 5% water capacity and 20 ± 2 °C.
(c) % Deviation from the Average = Standard Deviation/Average x 100

experiments conducted over a 4 year period using samples of soil from two agricultural plots will be examined.

In all experiments, soil samples were collected from the upper 10 cm of fields at random intervals between June 1981 and October 1985. Soil was not taken when it had been frozen or subjected to drought for more than 30 consecutive days. After collection, soils were brought into the laboratory, passed through a 2 mm sieve, and allowed to stand at room temperature (20 to 22°C) in a moist condition (40 to 45% water capacity) for 7 to 28 days. Thereafter, germinated seeds were removed, and the

Table 5   Experiments conducted in a single laboratory showing the mineralization of lucerne-grass-green-meal (0.5%) in samples of a sandy silt (a) collected over a 4 year period without regard to season or soil cropping. (b)

Date of Soil Collection	Carbon Mineralization: Cumulative Values (mg Carbon Dioxide/100 g dry wt Soil)			
	7	14	21	28
29 IX 83	224	290	333	375
29 XI 83	157	223	267	309
2 II 84	174	245	286	316
27 III 84	176	251	294	331
27 IV 85	173	270	334	379
11 VI 85	198	257	290	327
31 VII 85	224	312	372	416
11 VIII 85	237	304	355	393
21 X 85	243	326	384	442
Average ± Std Deviation	200 ± 32	269 ± 30	322 ± 41	365 ± 47
% Deviation from the Average (c)	± 16%	± 11%	± 11%	± 13%

(a) Total carbon = 2.6%; total nitrogen = 0.24%; pH = 5.4
(b) Soils incubated under aerobic conditions in the dark at 40 ± 5% water capacity and 20 ± 2 °C.
(c) % Deviation from the Average = Standard Deviation/Average x 100

soils were mixed with lucerne-grass-green- meal (0.5%) and incubated under aerobic conditions in the dark at 20 $\pm$ 2°C and 40 $\pm$ 5% water capacity. After 7, 14, 21 and 28 days, the quantities of carbon dioxide evolved from the soils were determined. Four replicates were used in each test.

In 15 tests with a loamy sand soil collected at random intervals between June 1981 and July 1985, the largest deviation from the average respiration value was $\pm$ 10% on Day 7 of the experiments (Table 4). After 14, 21 and 28 days of incubation, the deviations were constant at $\pm$ 7%.

For the 9 tests conducted with a loamy silt soil collected at random intervals between September 1983 and October 1985, the largest deviation was $\pm$ 16% and this was also on Day 7 of the experiments (Table 5). Thereafter, the deviations ranged from $\pm$ 11% to 13%. These data quite conclusively show that when soil respiration tests are conducted by a single laboratory, respiration values are quite reproducible. To be sure, there is some variation, but it is less than if different laboratories conduct the same test. It is of interest that there is no particular correlation between the date or season of collection and rates of mineralization.

Results with unamended soils

The objective of one set of recommended tests is to determine if pesticides disturb the microbial degradation of stabilized soil organic matter. In these tests, the respiration of pesticide-treated but otherwise unamended soil is measured.

Two of the above mentioned laboratories conducted such tests. For this purpose, samples of 2 different agricultural soils were treated with 100 and 200 mg mercuric chloride/kg dry wt soil and incubated under aerobic conditions in the dark at 20 $\pm$ 2°C and 40 $\pm$ 5% water capacity. Respiration measurements were made after 1, 2, 4, 7, 14, 21 and 28 days.

The results of the experiments are given in Table 6 as % stimulation or % inhibition of soil respiration in reference to the untreated control. In spite of the high doses of mercuric chloride, the reactions of the soils to the chemical were hardly measurable. The maximum deviation from the basal respiration of either of the soils was a stimulation of respiration which amounted to 14% more than that of the control (Table 6, soil 1, 100 mg/kg). When this data is compared to that in Tables 2 and 3, where initial inhibition of respiration in nutrient-amended soils reached as high as 98%, it becomes obvious that unamended soils are insensitive and by no means suited for testing for the side-effects of pesticides on the microbial mineralization of carbon in soils.

The stimulations noted in these experiments are of interest. They are often seen when toxic materials are added to unamended soils. In the case of mercuric chloride, which in itself does not provide a potential source of carbon for the microflora, stimulation of carbon mineralization is most probably due to killing of some microorganisms by the poison followed by mineralization of these freshly killed cells by the surviving microflora. In cases where overdoses of organic pesticides are applied, a part of the carbon dioxide might also come from the pesticide itself.

Table 6   Influence of 100 and 200 mg mercuric chloride on the respiration of two unamended soils incubated under aerobic conditions in the dark at 20 $\pm$ 2 °C and 40 $\pm$ 5% soil water capacity

Days of Incubation	% Stimulation (+ Values) or % Inhibition (- Values) of Soil Respiration			
	mg Mercuric Chloride/kg Dry wt Soil			
	100		200	
	Soil 1	Soil 2	Soil 1	Soil 2
1	+ 14	0	- 9	0
2	+ 10	+ 10	+ 6	+ 10
4	+ 7	+ 7	+ 7	+ 7
7	+ 7	0	+ 2	+ 5
14	+ 12	- 8	+ 2	0
21	+ 13	- 4	+ 4	- 2
28	+ 11	- 8	+ 4	- 5

SUGGESTION FOR IMPROVEMENT OF RECOMMENDED TESTS

Soils: the soils used in experiments should be broadly distributed and representative at an international level. If possible, only freshly collected soils should be used. If soil must be stored, it should be held in a moist condition at 2 to 4°C for not more than 3 months. A sandy soil of low organic content (less than 1% organic carbon) and a loamy soil of higher organic content (greater than 1.5% organic carbon) are suggested. The pH of the soils should be between 5.5 and 7.0.

Sequence of tests: With the expressed goals of protecting soil quality and using research capacities and facilities as rationally as possible, the recommended sequence of soil respiration tests shown in Appendix B is endorsed.

REFERENCES

Anderson, J.P.E., 1982, in Methods of Soil Analysis, Part 2, 2nd edition, edited by A.L. Page, R.H. Miller and D.R. Keeney (Madison, American Society of Agronomy) p. 831.

Greaves, M.P., Poole, N.J., Domsch, K.H., Jagnow, G. and Verstraete, W., 1980, Recommended Tests for Assessing the Side-effects of Pesticides on Soil Microflora, Technical Report, Agricultural Research Council, Weed Research Organisation, No. 59.

Greaves, M.P., Cooper, S.L., Davies, H.A., Marsh, J.A.P., and Wingfield, G.I., 1978, Methods of Analysis for Determining the Effects of Herbicides on Soil Microorganisms and their Activities, Technical Report, Agricultural Research Council, Weed Research Organisation, No. 45.

CHAPTER 6

RESPIRATION AND DEHYDROGENASE
AS SIDE-EFFECTS INDICATORS

H.-P. Malkomes

INTRODUCTION

At the workshop on "regulatory aspects of side effects of pesticides on the soil microflora" in November 1979 at Windsor (UK), which preceded the September 1985 meeting at Cambridge (UK), some routine tests were recommended to investigate side effects of pesticides on the soil microflora (Greaves et al., 1980). At that time in our institute another program was being tested to evaluate methods of investigating side effects of pesticides (Malkomes and Wohler, 1983, Malkomes and Pestemer, 1984), but the unfinished research did not enable us then to decide which of them would be useful. In some cases, however, the methods recommended at the 1979 workshop were unsatisfactory for us. The investigations we have done subsequently have confirmed some of our earlier observations. In the following text some of our results, published in the last 5 years, will be presented in a somewhat modified form. Only soil respiration and dehydrogenase activity (i.e. electron transport system or ETS) methods will be presented, while nitrogen transformation and others will be omitted.

Although, in many countries, the authorities responsible for the approval of pesticides are probably not obliged to follow the recommendations coming from these workshops, it would be useful to more or less adopt the different methods. Since 1979, disadvantages in and improvements to the then recommended tests have been established. Further possible improvements have been investigated by different research groups and a continued effort at evaluation and, if appropriate, improvement, by further workshops is needed.

MATERIALS AND METHODS

At some earlier workshops it was recommended that side-effect experiments should use 2 special soils which should be representative for many agricultural soils. In Table 1 the soils "BBA" and "Sickte" are such types, whereas the high humus "HB" soil is a less common type. All these soils were used in laboratory and field trials. Under field conditions, however, we must consider greater horizontal and vertical variation of the data of Table 1.

Laboratory trials

The herbicides used are shown in Table 2. Two of them are dinitrophenol containing preparations known for their side effects. To achieve "normal dosage" we simulated, as recommended in 1979, the practical field dosage distributed throughout the upper 5 cm soil layer. In addition, we used a dosage 10 times higher, as was also recommended. This higher concentration is possible under field conditions if we assume a penetration only into the uppermost 0.5 cm soil layer. All the preparations were mixed into the soils. The incubation for several weeks at 20°C and 60% m.w.h.c. also followed earlier recommendations. During the whole time the water content was stable. The microbiological investigations were done periodically depending on the special demands. In some cases a parallel trial was started using lucerne meal amendment (0.5 g/100 g soil). Three replicates were used.

Field trials

In the field all herbicides were applied at recommended dosages in the springtime. In some cases, dinoseb acetate (Aretit) was incorporated into the upper soil layer to give better comparability to laboratory conditions. In a parallel trial half of the areas were maintained without plant cover, while in the remaining areas wheat (at Sickte) or field peas (at BBA) were grown. All weeds were removed mechanically without disturbing the soil excessively. Soil samples were taken from 0-5 cm soil depth during the whole season and investigated in the laboratory. Under field conditions four replicates were used.

Table 1   Properties of the three test soils

Soil type	particle size distribution (%)			C_{org} (%)	pH (in 0.1N KCl)	Dry soil density (kg/cm^3)	max. water holding capacity (g/100 g soil)
	clay	silt	sand				
BBA ; loamy sand	1.0	26.3	72.7	0.8	5.6	1.37	23.7
HB ; very humous silty sand	3.4	48.3	48.3	13.5	7.2	0.57	106.4
Sickte ; clay loam	25.9	50.3	23.8	2.6	7.4	1.38	42.0

Table 2   Field dosages of the herbicides applied to test soils

Soil	Aretit flüssig (492 g dinoseb acetate/l)		NaTA (95% TCA)		Ro-Neet (720 g cycloate per l)		Wacker Murbetex 0 (29.4% propham + 14.55% medinoterb acetate)	
	l/ha	ul/kg soil(a)	kg/ha	mg/kg soil(a)	l/ha	ul/kg soil(a)	kg/ha	mg/kg soil(a)
BBA	4	5.3	30	39.7	4	5.3	7.5	9.9
HB	4	6.6	30	49.7	4	6.6	15.0	24.8
Sickte	4	4.9	30	36.8	4	4.9	12.0	14.7

(a) herbicidal concentration was related to 5 cm soil depth and a soil moisture of 60% m.w.h.c.

RESPIRATION AND DEHYDROGENASE AS SIDE-EFFECTS INDICATORS

Microbial investigations

Under our test conditions long term respiration was an indicator of carbon mineralization in the soil. As described by Malkomes and Halstrick (1985), the CO_2 evolution from the soil, over a period of 8 weeks after herbicide application, was measured by absorption in KOH and subsequent titration, using closed bottles purged with 1 litre CO_2 free air per hour. In the graph of data from the laboratory trials we show the cumulative CO_2 formation. As recommended in earlier symposia, we used 2 systems, soil alone and soil amended with lucerne meal. It was not practicable to measure long-term respiration in the field parallel to measurement of dehydrogenase activity and short-term respiration. Samples for these measurements were amended with straw meal (2 g/100 g soil) and NH_4NO_3 solution (2 ml of a 2.5% solution per 100 g soil) and were only incubated for 14 days. In the graphs showing long-term respiration in field trials we, therefore, present the cumulative CO_2 production only for this time.

In our investigations, short-term respiration was an indicator of the reaction of the microbial biomass, and is possibly comparable with the method of Anderson and Domsch (1978) to measure active microbial biomass. As described by Malkomes (1982a, 1986) the soil samples taken at different time after herbicide application were amended with glucose (0.1-2 g/100 g soil) followed by measurement of CO_2 formation during 48 hours using an URAS (infra-red) gas analyser. In the following graphs only the amount of CO_2 produced within the first 12 hours is shown. The the size of the respiration curves during the 48 hours period, especially the peak phase, gives further information on pesticide effects.

In many investigations we have found dehydrogenase activity to be a sensitive indicator of microbial activity in soil. Recently, Trevors et al. (1982) mentioned this enzymatic activity for studying microbial electron transport system (ETS) activity and its relation to O_2 consumption in agricultural soils. The method we used has been described by Malkomes (1981). Soil samples taken at different times after herbicide application in field and laboratory were investigated using TTC (triphenyl tetrazolium chloride) as an indicator during photometric measurement (Thalmann, 1968). In our trials the soil was not amended with any easily decomposible organic material.

In the following graphs normally the standard deviation was indicated.

85

RESULTS

Laboratory trials

    Using some examples we will present some pecul-
iarities of soil respiration measurements found in our
laboratory trials. At first the different activities in
soil without lucerne meal amendment will be compared.
Figure 1 shows the reaction of 3 different microbial

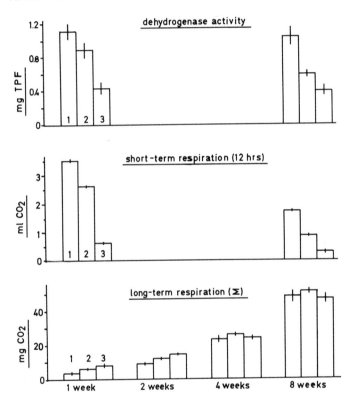

Figure 1   Comparison of the effect of a herbicidal mixture of propham
            + medinoterbacetate on 3 microbial activities in BBA soil
            without lucerne meal amendment under laboratory conditions;
            results per 100 g soil dry matter equivalent (data from
            Malkomes, 1984), 1 = control, 2 = propham +
            medinoterbacetate (1x), 3 = propham + medinoterbacetate
            (10x).

activities to a herbicide of the dinitrophenol type,
using a recommended loamy sand soil (BBA). With
long-term respiration, 2 phases can be observed within 8
weeks of incubation. In the first 2 weeks the cumulative

CO_2 formation was stimulated increasingly with increased dosage. Subsequently, this effect disappeared. When we first found this stimulatory effect many years ago, we were astonished, because until then only inhibitions had been discussed as being serious side effects. We now know that these effects are not real stimulations for other microbial analyses showed inhibitions. In Figure 1 those negative effects depending on increased dosages can be seen with short-term respiration and dehydrogenase activity over a

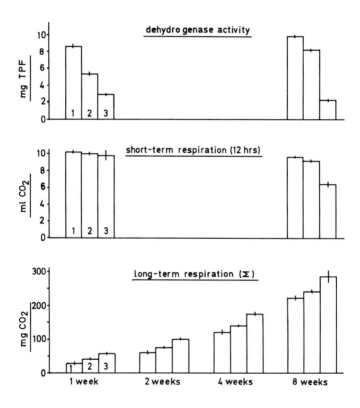

Figure 2   Comparison of the effect of the a mixture of propham + medinoterbacetate on 3 microbial activities in HB soil without lucerne meal amendment under laboratory conditions; results per 100 g soil dry matter equivalent (data from MALKOMES, 1984), 1 = control, 2 = propham + medinoterbacetate (1x), 3 = propham + medinoterbactate (10x)

period of 8 weeks. From our work, and from data of Schroder (1984) we assume, that biocidal pesticides disturb some parts of the microbial biomass and this effect is followed by increased mineralization of killed or damaged organisms and parts of the soil organic matter. This effect not only caused increase in long-term respiration in many cases, but also as we found in our earlier investigations (Malkomes, 1982b, 1985c, 1987

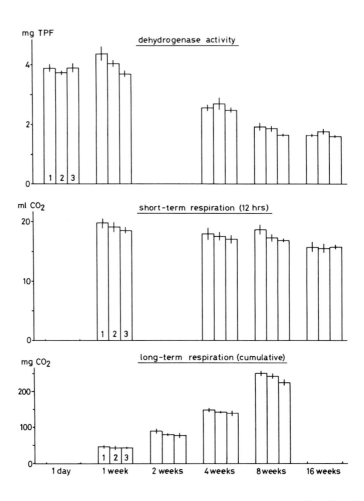

Figure 3 Comparison of the effect of TCA on 3 microbial activities in Stickte soil without lucerne meal amendment under laboratory conditions; results per 100 g soil dry matter equivalent (data from Malkomes, 1985b). 1 = control, 2 = TCA (1x), 3 = TCA (10x)

1987), stimulations of nitrogen mineralization.

The same dinitrophenol containing herbicide also was applied to a high humus soil (HB) using dosages adapted to its greater sorption capacity. Although this soil was quite different to the first one, and had a greater microbial activity, similar reactions of the micro-organisms were found (Figure 2). Cumulative CO_2 formation (long-term respiration) was also stimulated

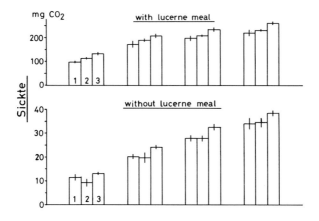

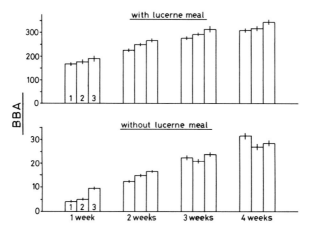

Figure 4   Comparison of the effect of cycloate on long-term respiration of 2 soils with and without lucerne meal amendment under laboratory conditions; results per 100 g soil dry matter equivalent (data from Malkomes and Halstrick, 1985). 1 = control, 2 - cycloate (1x), 3 = cycloate (10x).

more with increased dosage, although in the soil the effect continued over the whole period of 8 weeks. As before, short-term respiration and dehydrogenase activity were inhibited more strongly by the higher herbicide dosage. The absolute extent of the effect was mostly greater when compared to that in the sandy loam soil (BBA), but the relative effects were similar.

The reactions of the soil microorganisms shown in Figures 1 and 2 were common in many of our investigations of the effects of herbicide applications. Not all preparations, however, caused similar effects. TCA applied to a clay loam soil (Sickte) provides such an exception as shown in Figure 3. Here cumulative CO_2 formation (long-term respiration) was inhibited more when the dosage increased, this effect continuing over the whole period of 8 weeks. Thus the effect was opposite to the examples in Figures 1 and 2. Short-term respiration was also inhibited for at least some time, but the inhibition was weaker. The reaction of dehydrogenase activity was similar, too, although the effects observed were mostly only weak.

The methods recommended in 1979 (Greaves et al., 1980) also mentioned a respiration test using soil amended with lucerne meal. In our investigations we have often used this variation, too. Some of our results are presented in Figure 4 for this long-term respiration after the application of the herbicide cycloate to 2 standard soils. After lucerne meal amendment, the cumulative CO_2 formation of both soils rose strongly. In the first weeks about 10 times the normal level was reached, especially in the lighter soil (BBA). Using both soils without lucerne meal different herbicidal effects depending on the soil type were found. In the loamy sand soil (BBA) greater stimulations occurred with increased herbicide dosage for 2 weeks, whereas in the clay loam soil (Sickte) only the higher dosage stimulated. This effect here continued over the whole period of 4 weeks. If lucerne meal was added to the soils a different reaction to the herbicide was found. In both soils, stronger stimulations occurred with higher herbicide dosage, an effect similar to that already observed in the clay loam soil (Sickte), higher in humus content but without lucerne meal.

When the experiment reported in Figures 1 and 2 was repeated but amended with lucerne meal, long-term respiration was more stimulated with increasing herbicide dosages, too. However, the inhibitory effects on short-term respiration and dehydrogenase activity were relatively diminished for both herbicide dosages in the lighter soil (BBA), whereas in the very humous soil similar reactions occurred with and without lucerne meal.

Field trials

In Figure 5 the results of a trial of the influence of the dinitrophenol herbicide dinoseb on 3 microbial activities in loamy sand soil (BBA) with and without pea plants are shown. The long-term respiration showed no reaction to the herbicide. In the field trials, unlike laboratory trials, long-term respiration was only measured as accumulated CO_2 production for 2 weeks from soil samples taken at intervals as for the other microbial activities. Short term respiration, however, was inhibited by the normal field dosage from the 49th to the 78th day in soil with and without plants, whereas dehydrogenase activity was inhibited mainly in soil with plant cover. In 1981, a parallel trial was run using the same area, but long-term respiration was omitted. Here dehydrogenase activity was more or less inhibited by the herbicide in the soil with and without pea plants. Short-term respiration, however, was not markedly affected.

Similar trials to those mentioned above were run using a clay loam soil (Sickte). In 1980, the herbicide sometimes caused weak stimulations of long-term respiration in the soil with and without wheat plants, which were similar but less marked than in our laboratory trials with another dinitrophenol containing herbicide in another soil (Figure 2). The short-term respiration, however, was sometimes only slightly inhibited in the soil without plants. Dehydrogenase activity at least at 3 sampling periods showed again more or less inhibitions in the soil with and without plants due to the herbicide. When this trial was repeated in 1981 similar results were found for dehydrogenase activity and short-term respiration (Figure 6). Long-term respiration was omitted from this trial.

Comparison of laboratory and field trials

In Figure 7 the relative effect of the herbicide dinoseb on the sensitive and indicator microbial activities "short-term respiration" and dehydrogenase activity in soils from laboratory and field trials were

91

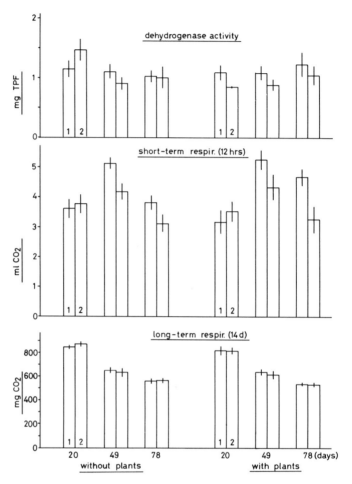

Figure 5   Comparison of the effect of dinoseb on 3 microbial activities in BBA soil under field conditions with and without pea plants, 1980; results per 100 g soil dry matter equivalent (data from Malkomes and Pestemer, 1984).
1 = control, 2 = dinoseb (1x)

compared. In the loamy sand soil (BBA) greater inhibitions were found under laboratory conditions than in the field. In the clay loam soil (Sickte) no great differences occurred between laboratory and field. With this soil, however, the herbicide caused marked inhibitions of the dehydrogenase activity only under laboratory conditions. Under field conditions, similar

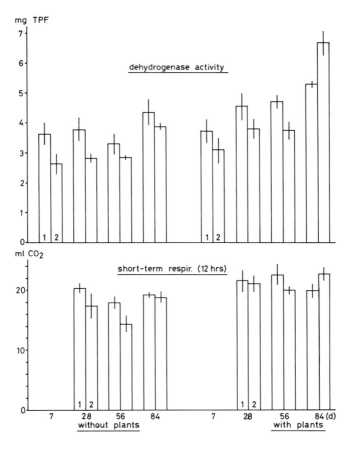

Figure 6   Comparison of the effect of dinoseb on 2 microbial activities in Sickte soil under field conditions with and without wheat plants, 1981; results per 100 g soil dry matter equivalent (data from Malkomes and Pestemer, 1984). 1 = control, 2 = dinoseb (1x)

reactions occurred but, in 1981, short-term respiration was also slightly inhibited.

DISCUSSION

Our investigations using different herbicides, have shown that short-term respiration, dehydrogenase activity and cumulative long-term respiration differed markedly in their reaction to anthropogenic chemicals. Under laboratory conditions this was particularly clear. Normally, dehydrogenase activity and short-term resp-

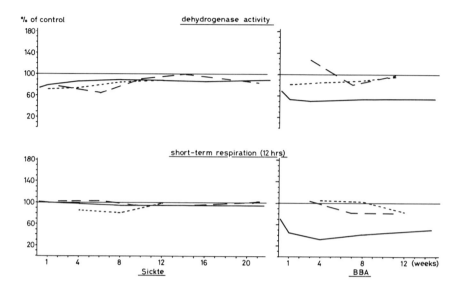

Figure 7   Comparison of the effects of the normal dosage of dinoseb on
              2 microbial activities in 2 soils under laboratory and field
              conditions (data from Malkomes & Wohlor, 1983 and Malkomes &
              Pestemer, 1984). _____ = laboratory 1981
              ___ ___ = field 1980   - - - - = field 1981

iration, the latter being used in a modified form by
Anderson and Domsch (1978) to determine active microbial
biomass, were more inhibited with increased dosages.
Even with normal dosages, inhibitory effects occurred
with some pesticides. The cumulative long-term
respiration, however, showed greater stimulations with
increased dosage of some herbicides, but with others an
inhibition occurred under certain circumstances. This
stimulatory effect was used by Jenkinson and Powlson
(1976), in a modified form, to determine microbial
biomass.

       In view of these effects, therefore, we recommend
short-term respiration, or as its equal substitute
dehydrogenase activity, for routine tests with
pesticides. They enable us to determine in a very
sensitive way the microbial activity of soil samples from
field and laboratory trials within 1-2 days. If both
activities are compared we prefer dehydrogenase activity
to short-term respiration in our investigations. It is

94

not restricted to a narrow instrumental capacity and its sensitivity towards pesticides is similar or sometimes even better. In recent years many authors have included this enzymatic activity in their investigations.

If long-term respiration is used for routine tests on side effects of pesticides on soil microorganisms, additional investigations of further microbial parameters are necessary to allow adequate interpretation of its effects. Furthermore another disadvantage of this method is evident. Long-term respiration measurements normally cannot be carried out in the field using the technique recommended for use in laboratory tests (Greaves et al., 1980). In many cases, however, investigations of the microbial activity after application of pesticides in the field are useful. With long-term respiration a valid comparison of laboratory and field trials is virtually impossible.

For the interpretation of all results of side effect trials an evaluation scheme is needed. Because we often found stimulatory effects resulting from heavy pesticide treatment in our investigations, especially with long-term respiration and also with nitrogen mineralization, we have modified the interpretation model of Domsch et al., (1983). With our modified model (Malkomes, 1985) the interpretation of positive as well as negative deviations from the control soil is now possible. In addition, we have shortened the time over which effects are allowed to occur. The length of this period was adapted to the practical situation in agriculture, where a new application of pesticides is not uncommon in the same growing season, e.g. after about 3 months.

REFERENCES

Anderson, J.P.E and Domsch, K.H., 1978, Soil Biology Biochemistry, 10, 215.

Domsch, K.H., Jagnow, G. and Anderson, T-H, 1983, Residue Reviews, 86, 65.

Greaves, M.P., Poole, N.J., Domsch, K.H:, Jagnow, G. and Verstraete, W., 1980, Recommended Tests for Assessing the Side-effects of Pesticides on the Soil Microflora - Technical Report. Agricultural Research Weed Council, Weed Research Organisation No. 59.

Jenkinson, D.S. and Powlson, D.S., 1976, Soil Biology Biochemistry 8, 290.

Malkomes, H-P., 1981, Zentralblatt für Bakteriologie. 2. Abteilung, 136, 451.

Malkomes, H-P., 1982a, Zentralblatt für Mikrobiologie, 137, 97.

Malkomes, H-P., 1982b, Zentralblatt für Mikrobiologie, 137, 525.

Malkomes, H-P., 1984, Zentralblatt für Mikrobiologie, 139, 441.

Malkomes, H-P., 1985a, Berichte über Landwirtschaft, Sonderheft, 198, 134.

Malkomes, H-P., 1985b, Les Colloques de l'INRA, 31, 191.

Malkomes, H-P., 1985c, Z. Pflanzenkr. Pflanzensch, 92, 489.

Malkomes, H-P., 1986, Nachichtenbl. Deut, Pflanzenschutz, 38, 113.

Malkomes, H-P., 1987, Zentralblatt für Mikrobiologie, in press.

Malkomes, H-P., and Halstick, S., 1985, Zentralblatt für Microbiologie, 140, 381.

Malkomes, H-P., and Pestemer, W., 1984, Zeitschrift für Pflanzenkrankenheiten und Pflanzenschutz, Sonderh. X, 193.

Malkomes, H-P., and Wohler, B., 1983, Ectotoxicology and Environmental Safety, 7, 284.

Schroder, M., 1984, Dissertation University of Hannover, p 163.

Thalmann, A., 1986, Landw. Forsch. 21, 249.

Trevors, J.T., Mayfield, C.I, and Inniss, W.E., 1982 Microbiol Ecology, 8, 163.

CHAPTER 7

NITROGEN TRANSFORMATION:
PRACTICAL ASPECTS OF LABORATORY TESTING

C. R. Leake & D. J. Arnold

INTRODUCTION

The processes of ammonification and nitrification are parts of the nitrogen-mineralisation cycle in soil providing nitrogen in a utilisable form for plants.

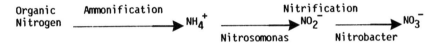

Figure 1   Nitrogen transformations in soil

During ammonification, ammonium is released from organic matter by the activity of a wide variety of micro-organisms (bacteria, actinomycetes and fungi) in the soil (Figure 1). In contrast, the oxidation of ammonium through nitrite to nitrate is carried out by a relatively restricted group of microorganisms, the most well known being Nitrosomonas and Nitrobacter. The effects of a pesticide on these two interlinked processes are determined by measurement of the changes in the levels of ammonium, nitrite and nitrate-nitrogen in the soil.

Determining the influence of a novel pesticide on soil nitrogen transformations in the field poses many problems. It is complicated by the need to assess the relative importance of any perturbations arising from external influences such as severe water stress, temperature, soil type, cropping practices and perhaps most significantly in this case, the addition of organic or inorganic nitrogen fertilisers. It is possibly because of such complex interactions that researchers have tended to bring the field into the laboratory, in order to have at least some control over these parameters.

Despite the inherent natural variability of soil microbial activity some government registration authorities require pesticide manufacturers to produce a measurement of the effect of a chemical on nitrogen transformations. The need arises therefore, to continue to search for the most reliable and reproducible laboratory methods. The methodology, however, should enable us to be confident about the chemical's behaviour in the field and current laboratory studies are not necessarily satisfying that criterion.

The object of this paper is to outline some of the practical considerations in tackling nitrogen transformation studies in the laboratory. Although many of the points may seem rather obvious, it is clear that too little attention is paid to the many parameters which undoubtedly influence the results and consequently the interpretation of a particular study.

EXPERIMENTAL

Experimental design

Table 1 summarises the experimental layout for two studies which have been carried out using a herbicide, benazolin–ethyl, and an acaricide, clofentezine. The layout broadly follows the scheme recommended by Greaves et al. (1980). The pesticide is generally applied to soil samples in a solvent and therefore a solvent control was used to investigate any effect the solvent may have on nitrifiers. Clofentezine, for example, is extremely insoluble in water and of low solubility in many common organic solvents such as acetone, therefore, dichloromethane was used in both control and treated samples. Organic nitrogen in the form of powdered lucerne meal, is often added to supplement the inherent soil organic nitrogen in order to demonstrate that ammonification is taking place. The uniformity of mixing the substrate with the soil will inevitably influence the rate of release of ammonium as a function of the relative contact of the substrate with the soil matrix.

Soil handling and treatment

Two contrasting soil types were selected, a loamy sand and a clay. These soils are from sites which had been free from pesticide treatments for many years in

Table 1   The experimental design used to study the effect of
a pesticide on soil nitrogen transformations

TREATMENT	SOIL:LOAMY SAND	SOIL:CLAY
Untreated soil (solvent control)	✓	✓
Untreated soil (amended with lucerne) solvent control	✓	✓
Soil treated at field rate (no lucerne)	✓	✓
Soil treated at field rate (amended with lucerne)	✓	✓
Soil treated at 10 x field rate (amended with lucerne)	✓	✓

All measurements carried out in duplicate.

order to reduce the possibility of pre-activated metabolic pathways.However, their major significance is that soils from the same site are used in metabolism studies in our laboratory and thus provide a link in terms of pesticide breakdown and microbial activity. The soils were sampled from the field to a depth of 15 cm. and used as soon as possible after collection, but certainly within 14 days. Fresh soils are essential as highly variable microbial activity has been observed, for example, with the West German Speyer soils which are often stored for long periods prior to despatch to the recipient. Indeed, several workers have shown that the overall microbial activity of Speyer soils is considerably less than that of a corresponding soil sampled freshly from the field.

Pesticide treatment rates should be realistic in terms of field rate. Hence, the maximum recommended field rate should be used with a rate of ten times that level to provide a worst case situation. Table 2 (abstracted from Domsch and Paul, 1974) summarises the effect of selected herbicides on N-mineralisation and shows that some claims of an effect on nitrogen transformations are based on excessively high treatment rates.

Perhaps one of the most important aspects of any laboratory experiment is the provision of sufficient replication in order to provide confidence in the data produced. Although some laboratories use duplicate treatments in nitrogen-transformation studies, as we have done, our experience has convinced us that this is insufficient and that 3-4 replicates are needed to give adequate reproducibility. Some workers analyse sub-samples from a single soil pot treatment, however this is not a satisfactory method as it can lead to errors due to non-homogeneity between sub-samples. This conclusion is based on our experience of sub-sampling soils treated with radiolabelled pesticides. Analysis of separately treated aliquots is preferred.

Table 2   Effects of selected herbicides on N-mineralisation in soil

Author	Chemical	Treatment	Effect
Chunderova and Zubets	Chlorpropham	high rate	-
		field rate	0
Helweg	Chlorthiamid	500 kg/ha	-
Bioko et al	2,4-D	0.75-7.5 kg/ha	0
Chunderova and Zubets	Dalapon	field rate	0
Van Schreven et al	Dalapon	X10 field rate	-
Chandra	Diallate		-
Sommer	Diallate		0
Debona and Audus	Dichlobenil	500 kg/hg	-
Domsch and Paul	Diuron	variable rates	-
Debona and Audus	Ioxynil	X100 field rate	-
Van Schreven et al	Mecoprop	field rate	0
		X10 field rate	-
Tu and Bollen	Paraquat	field rate	0
Anderson and Drew	Paraquat	field rate and	
		X10 rate	0 or +
Debona and Audus	Paraquat	X100 field rate	-
Tena et al	Phenmedipham	50-500 mg/g soil	-
Sommer	Phenmedipham	field rates	0
Domsch and Paul	Simazine	field rates	0

0 = No change in mineralisation   - = Depression   + = Increase
From Domsch and Paul, 1974

INCUBATION

Incubation Equipment

There is a multitude of systems used within laboratories to incubate soils with several variations on each design (Figure 2). Although the effect of a pesticide on ammonification and nitrification is determined from a relative comparison between treated and untreated soil, different systems are likely to enhance or deplete different controlling factors. Some workers use open pots, perhaps within a plastic bag to reduce moisture losses. Other possibilities include open or closed columns with packed soil or intact soil cores. The perfusion system is commonly used and is described by Laskowski and Bidlack (1977) with some versions incorporating a peristaltic pump to circulate the perfusate. In comparison with the conical flask method, where there is no removal of nitrate as it accumulates with time, the perfusion technique removes it by leaching and then recycles it. Because the majority of laboratory experiments are conducted in enclosed aerobic systems there is very little opportunity for denitrification. Hence, significant accumulation of nitrate may occur over a 6-10 week incubation period.

Moisture content and flask geometry

It is essential to maintain a constant soil moisture content and the technique used in our studies is similar to the system used for our radiolabelled pesticide metabolism studies. A supply of air is passed through a bubbler tube dipped into water to ensure a moist air input and the expelled air is removed from just above the soil surface, (Figure 3). This method has been successful in controlling the moisture content of a variety of soils at 40% water holding capacity with virtually no loss over a six week period. Clearly the moisture content can influence the rates of ammonification and nitrification and most workers use between 40% and 60% moisture holding capacity.

Radiolabelled compounds have shown that the geometric shape of the incubation flask and the soil void:volume ratio can produce different rates of pesticide degradation (Blum et al., 1983) and there is every reason to suspect that a similar effect might be observed on the ammonification and nitrification processses.

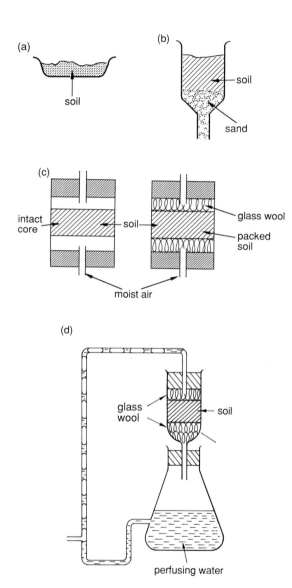

Figure 2   Incubation systems : (a) open pots (b) open columns
(c) closed columns with intact and packed cores
(d) perfusion system

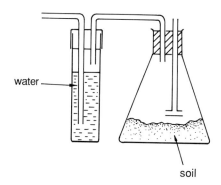

water

soil

Figure 3   Moisturising bubbler and flask.

Temperature

Most laboratory studies are carried out between 20 and 30°C probably because enhanced microbial activity at these temperatures results in an increased rate of nitrogen transformation. However, Anderson et al. (1964) and Marsh et al. (1979) have considered the effect of temperature and have shown that at lower temperatures rates of nitrification were much lower than previously thought. It is not clear what effect a pesticide might have on nitrification at lower temperatures, except the influence on rate of degradation.

Treatment and calculation of treatment rate

Most field studies involve spraying the chemical onto the soil or plant. In contrast, however, in many laboratory studies, the pesticide is incorporated into the soil. In general, most literature reports the application rate in terms of kg chemical/ha relative to field rate. Some do not give the percentage of active ingredient when a formulation is used and most fail to state the assumptions made when converting this figure to concentration in a laboratory flask. In the case of a non-mobile pesticide the penetration into the soil in the field is unlikely to be greater than 3 cm though many workers assume 10 or 12.5cm. This in itself gives rise to a four fold difference in assumed concentration - a further two fold difference being possible according to the figure taken for assumed soil bulk density.

Metabolic degradation

An observed effect on nitrogen transformation processes may be due to the applied parent compound, however, because of metabolic degradation a response may be due also to the presence of a rapidly formed metabolite. An example of rapidly formed metabolites is the hydrolysis of esters to their corresponding acids such as occurs with the herbicide benazolin-ethyl. Benazolin-ethyl is degraded to the corresponding acid (benazolin) with a half-life of approximately 3 days.

EXTRACTION METHODS

Extraction

Because inorganic nitrogen accounts for only a small proportion (<2%) of soil nitrogen and is present in relatively transient forms, extraction and subsequent quantification are perhaps not as straight forward as might first appear.

A limited amount of work has been conducted on the efficiency of extraction of ammonium, nitrite and nitrate from soils. Many workers would say that, when studying the effects of a pesticide, it is only relative differences that are being studied, and therefore, extraction efficiency is not important. However, since relative extraction efficiencies can vary considerably, greater differences between treatments are likely to be recorded when a more efficient extractant is used. A recent study (Nieto et al., 1985) (Table 3.) has shown very large differences in nitrate extracted according to the type of extractant used. Since nitrite and nitrate are small anionic molecules with low specific anion adsorption it can be assumed that they are readily extracted from soils. Although workers generally use 2M potassium chloride, calcium acetate has been shown to release at least twice as much nitrate-nitrogen compared with other extractants. It must be said too, that although 2ml of extractant per 1g soil is commonly used, there are probably different soil solution:void volume ratio's within the extracting vessels and therefore probably different extraction efficiencies.

Freezing/thawing

In order to cope with a large number of samples, each to be analysed for ammonium, nitrite and nitrate-

Table 3   Comparison of reagents in extracting inorganic anions from Kimberling soil as determined by ion chromatography

Extractant	Nitrite-N	Nitrate-N
	μg/g soil	
Water	1.04	15.5
LiCl	0.93	16.9
KCl	0.92	13.8
Ca (C2H3O2)2	1.10	32.7
CaCl2	0.98	15.4

From Nieto et al., 1985.

nitrogen, soil samples are often frozen prior to extraction. This has been shown to cause a marked increase in amounts of inorganic nitrogen compared with the levels analysed directly e.g. Table 4 (Allen et al., 1962). In our work (see Figures 4 and 5) the effect was particularly dramatic especially in the early part of the studies, however, other workers have shown no effect (Gasser, 1985; Nelson et al., 1972). Although this effect may be related to soil type and the treatment given to the soil when it is brought into the laboratory, both our experiments showed the same effect irrespective of the pesticide treatment or soil type. It is a factor that needs further examination.

Table 4   The increase in Ammonium-Nitrogen due to freezing

Ammonium-Nitrogen, ppm oven dry soil			
Fresh soil		Increase during	Freezing of air
Mean	S.E.	Freezing	Air drying	dry soil
2.5	0.2	19.0	3.3	14.0

From Allen et al., 1962.

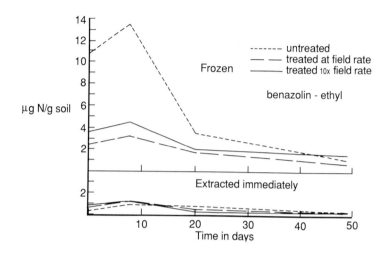

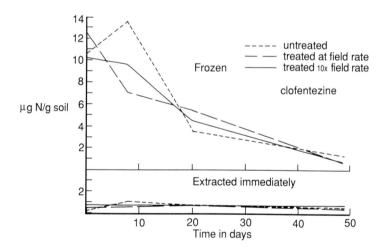

Figure 4 The effect of freezing/thawing on extractable ammonium-nitrogen from a sandy-loam soil content compared with immediate extraction

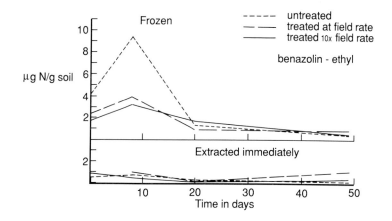

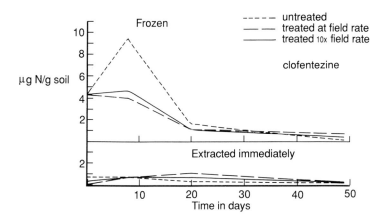

Figure 5   The effect of freezing/thawing on extractable ammonium-nitrogen content from a clay soil compared with immediate extraction

Biocides

Another area in which the methodology is in doubt is the stability of the extracts after extraction, prior to analysis. Most workers recommend analysis on the same day as extraction and some claim that the addition of chemical biocides such as chloroform or toluene prevent nitrification taking place in vitro. However, there is conflicting evidence as to their effectiveness (Robinson, 1967; Malhi, 1983; Lewis, 1961; Davies, 1940; Cunningham, 1962).

ANALYSIS

The study of the two processes, ammonification and nitrification requires accurate quantification of the three forms of inorganic nitrogen:-ammonium, nitrite and nitrate. Several methods are available but each has its own drawbacks and a degree of familiarity is necessary with each. What is sought is a sensitive method capable of handling a large number of samples fairly rapidly. At present, it is probably true to say that most workers use the auto analyser method as described by Greaves et al. (1978) or a modified version.

Auto analyser method

This method does have some advantages, as the colorimetric reactions are very specific. The use of a dialyser also assists clean up of the soil extract after filtration. It does, however, have the disadvantage of having to analyse each extract twice in order to determine nitrite and then nitrate by difference. Ammonium-nitrogen is determined by a fairly specific reaction with phenate and hypochlorite at high pH to form indophenol blue (Figure 6). Determination of nitrate and nitrite is by reduction of nitrate to nitrite and the formation of a red azo dye (Figure 6). The reduction used to be carried out in a zinc column, however this was subject to deterioration and has been replaced by the copper sulphate/hydrazinium sulphate method. The colour formation is produced by reaction of diazonium salt which then couples to form an azo-dye. The limit of detection for ammonium is good at around 8µg/l; although, at concentrations around 500 µg/l, Beer's law breaks down and it is, therefore, sometimes necessary to dilute the samples. The limit of detection for nitrite nitrogen is again good at around 10µg/l. Table 5 shows some reported limits of detection for various methods and should be regarded as an indicator of the approximate limits of detection depending on each worker's individual system. A good review of this topic is given by Keeney et al. (1982).

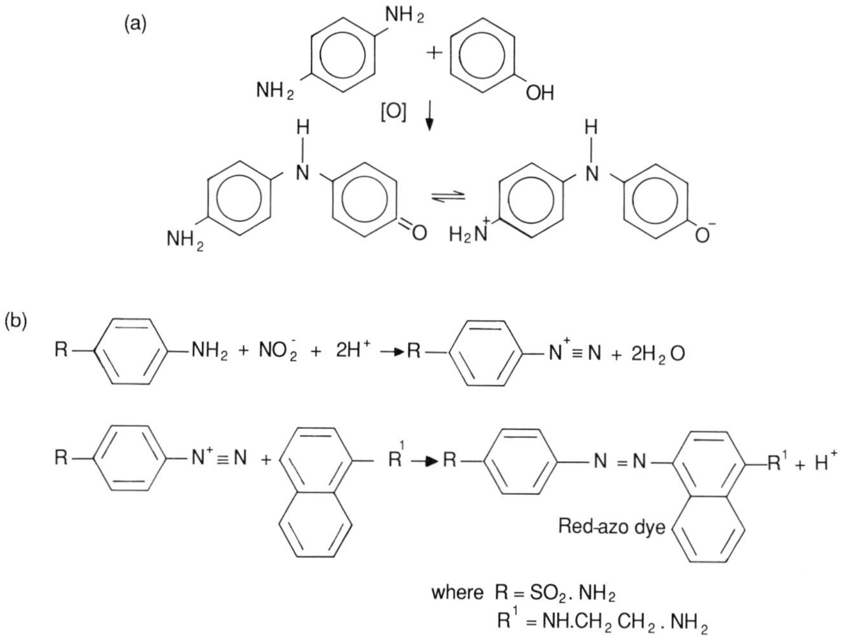

(a)

(b)

where $R = SO_2.NH_2$
$R^1 = NH.CH_2CH_2.NH_2$

Figure 6   Colormetric determination of NH_4^+ and NO_2^- (a) Indophenol blue and (b) modified Griess-Ilosvay red dye method

Specific ion electrodes

Whilst in theory this approach is straight forward, cheap and rapid, it is not widely used primarily due to problems with interferences during measurements.

Ammonium electrodes : Ammonium is converted to ammonia at pH's greater than 11. Several workers have

Table 5   Approximate limits of detection according to method

METHOD	ION		
	NH_4^+	NO_2^-	NO_3^-
Autoanalyser Indophenol blue/ Griess Ilosvay	8(a)	10	10
Specific Ion Electrode	500	–	1400
Ion Chromatography	–	14	3
Conductimetric	–	1000	25

(a) Figures are ug/litre

shown that reproducible results can be obtained fairly easily and these results are in close agreement with steam distillation methods. Interferences by mercury and amines have been reported but normally these should not be a serious problem in soils (Banwert et al., 1972). It is important to ensure that calibration curves are prepared in the soil extracting solution (e.g. potassium chloride) as they differ significantly from water. The limit of detection is reported to be about 500 ug/litre i.e. 50 times less sensitive compared with the autoanalyser method.

Nitrate eletrodes : The use of nitrate electrodes is quite a different story. They, in their present stage of development, require continual re-standardization and are subject to numerous interferences. There are also problems in trying to analyse in solutions like 2M potassium chloride as the chloride ion interferes with the electrode selectivity. However, many workers have reported some good comparisons, but recognition of possible serious interferences is necessary (Milham et al., 1970). Nitrate electrodes may have a place where a high degree of accuracy is not warranted.

The limit of detection is about 1400 ug/litre which is about 140 times greater than the autoanalyser method.

Nitrite eletrodes : A nitrite electrode has not

really been developed for analysis of soil samples, probably because of the serious interference problems and also because normally the quantity of nitrite in soil is normally very low it would not be within the likely limit of detection of an electrode. (Tabatabai, 1974).

Ion chromatography

The separation of ammonium, nitrite and nitrate using ion exchange columns has been reported recently in literature from several companies and the scientific press. The processes use buffer solutions under pressure from an HPLC pump (Thayer et al., 1980). Various types of anion exchange packing materials have been used as well as several types of detector including ultraviolet (at 200-120 mm) and electrochemical (Davenport et al., 1974). It is, of course, normally only possible to detect the anions nitrite and nitrate.

Separation of nitrate and nitrite on a cellulose anion exchanger can be achieved in about 10 min. and detected in a spectrophotometer with limits of detection of around 3 µg/litre for nitrate and 14 µg/litre for nitrite which is similar to the autoanalyser method (Gerritse, 1979).

The advantage of this method is that it could readily be automated with auto injector and integrator etc. However to determine ammonium as well could prove a little difficult.

CONCLUSIONS

A large number of variables have been shown to influence laboratory data on nitrogen transformations. Clearly before reproducible and reliable results are obtainable the techniques and methodology require some rigorous work in order to obtain a degree of accuracy that enables confidence in the experimental data.

Even if reliable methodology is developed it has not been established that data produced from laboratory experiments really reflect the situation in the field. In addition, if a stimulatory or inhibitory effect is found in both laboratory and the field it is unclear whether this would be of ecological significance in the enviroment when compared with natural perturbations.

REFERENCES

Allen, S.E. and Grimshaw, K.M., 1962, Journal of Science

Food and Agriculture, 13, 525.

Anderson, O.E. and Boswell, F.C., 1964, Soil Science Society American Proceedings, 28, 525.

Banwert, W.L., Tabatabai, M.A. and Bremner, I.M., 1972, Communication in Soil Science and Plant Analysis, 3, 499.

Blum, J.E., Stevenson, C.A., and Stainken, D.M., 1983, Environmental Toxicology and Chemistry, 2, 141.

Cunningham, R.K., 1962, Journal of Agricultural Science, 59, 257.

Davenport, R.J. and Johnson, D.C., 1974, Analytical Chemistry, 46, 13, 1971.

Davies, E.B., Coup, M.R., Thompson, F.B. and Hansen, R.P., 1940, The New Zealand Journal of Science and Technology, 348.

Domsch, K.H. and Paul, W., 1974, Archives of Microbiology, 97, 283.

Gasser, J.K.R., 1985, Nature, 181, 1334.

Gerritse, R.G., 1979, Journal of Chromatography, 171, 527.

Greaves, M.P., Cooper, S.L., Davies, H.A., Marsh, J.A.P., and Wingfield, G.I., 1978, Methods of Analysis for Determining the Effects of Herbicides on Soil Microorganisms and their Activities, Technical Report, Agricultural Research Council, Weed Research Organisation No. 45.

Greaves, M.P., Poole, N.J., Domsch, K.H., Jagnow, G., and Verstraete, W., 1980, Recommended Tests for Assessing the Side-effects of Pesticides on Soil Microflora, Technical Report, Agricultural Research Council, Weed Research Organisation, No. 59.

Keeney, D.R. and Nelson, D.W., 1982, in Methods of Soil Analysis Part 2. edited by A.L. Page, R.H. Miller and D.R. Keeney (Madison, American Society of Agronomy) p. 643.

Laskowski, D.A. and Bidlack, H.D., 1977, Down to Earth, 33, 12.

Lewis, D.G., 1961, *Journal of Science Food and Agriculture*, 12, 735.

Malhi, S.S., and Nyborg, M., 1983, *Soil Biology and Biochemistry*, 15, 581.

Marsh, J.A.P. and Greaves, M.P., 1979, *Soil Biology and Biochemistry*, 11, 279.

Milham, P.J., Award, A.S., Paul R.E., and Bull, J.H., 1970, *Analyst 1970*, 95, 751.

Nelson, D.W., and Bremner, J.M., 1972, *Agronomy Journal*, 64, 197.

Nieto, K.F. and Frankenberger Jr. W.T., 1985, *Soil Science Society America Journal*, 49, 525.

Robinson, J.B.D., 1967, *Plant and Soil*, 27, 53.

Tabatabai, M.A., 1974, *Communication in Soil Science and Plant Analysis*, 5, 569.

Thayer, J.R. and Huffaker, J.C., 1980, *Analytical Biochemistry*, 102, 110.

CHAPTER 8

NITROGEN TRANSFORMATION. RING TEST RESULTS

R. T. Hamm & N. Taubel

INTRODUCTION

The inorganic nitrogen in all soils is predominantly ammonium and nitrate. Several other inorganic forms have been postulated as intermediates of microbial transformation products of nitrogen, but have never been detected. In the following scheme (Figure 1) the importance of both these compounds in the total nitrogen cycle can be seen. It must be clearly emphasized that the following refers to about 2% of the total nitrogen in the soil and excludes the unexchangeable ammonia.

Ammonification and nitrification, as parts of the nitrogen-mineralization cycle, are useful criteria for studying possible side effects of pesticides. The enzymatic release of ammonia from organically-bound nitrogen and its immediate transformation to nitrate take place in all soils. The only stipulation is that aerobic conditions must be maintained to avoid the possibility of denitrification.

Preliminary trials are essential to determine whether a soil is suitable for studying the side effects of pesticides. Recently we have carried out extensive trials with different soils to find the answer to this question. Furthermore, several German chemical companies with long experience in the pesticide field have established a ring trial to check whether different analytical methods are suited to determine the rate of ammonification and nitrification in a soil. (See also Chapter 5). The companies and research workers that participated in the ring experiment, listed alphabetically, were BASF AG (R. Hamm), Bayer AG (J.P.E. Anderson and G. Hermann), Celamerck (D. Eichler),

Ciba–Geigy (J.A. Guth) and Hoechst AG (N. Taubel).

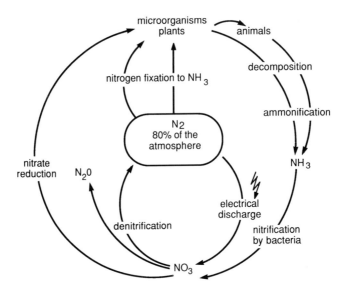

Figure 1   The nitrogen cycle

(Fixed NH4+ has been defined as ammonia, which cannot be replaced by a neutral K salt solution, e.g. 1 N K2SO4, 2 N KCl).

RESULTS

German standard soils

In Germany the BBA has issued guidelines in which it recommends the use of so called standard soils from the LUFA Speyer (Merkblatt No. 36 + 37) with characteristics as listed in Table 1, for degradation/leaching studies.

Table 1   German standard soils from Lufa Speyer
           (according to guidelines no. 36 and 37)

Drei Bodenarten, die bei der Landwirtschaftlichen Untersuchungs-
und Forschungsanstalt in Speyer bezogen werden sollten:

2 Boden (soils)	Org. C (%)	pH
2.1 Schwach humoser Sand (sand)	0.25–0.75	5.5–7.5
2.2 Stark humoser, lehmiger Sand (loamy sand)	2–3	5.5–7.5
2.3 Schwach humoser, sandiger Lehm (sandy loam)	0.5–1.5	5.5–7.5

The advantage of working with standard soils should
be that independent of the delivery date, the results
obtained with one type should be the same, or at least
comparable.   In Figure 2 it is demonstrated that the
unalterable pre-condition for a recommendation of   standard

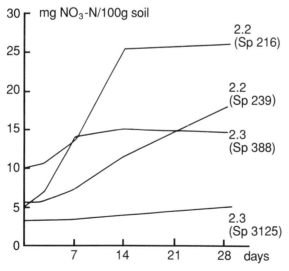

Figure 2   Nitrification rate in German standard soils (LUFA Speyer)

test soils, namely a comformable behaviour of a soil with the same characteristic but different delivery data, does not exist.

Pre-trials

These results show that a standard soil cannot be recommended in each case for nitrogen transformation studies. The description of a soil by its physico-chemical characteristics does not suffice to predict its biological behaviour. We are of the opinion that a measure of the nitrification rate is a much better criterion for the selection of soils. Nitrification studies with fresh soils from varying locations show (Table 2) that after the addition of 10 mg nitrogen as ammonium sulfate, ammonium has been transformed to nitrate within 28 days.

Table 2   Nitrification rate in several fresh soils

| No. | Soil origin | Soil texture | mg/100 g after | | | Suitability for N-trans-formation studies |
			0 days NO_3-N	14 days NO_3-N	28 days NO_3-N	
1	Kelsterbach	loamy sand	1.3	11.9	11.9	suitable
2	Hattersheim	clay	1.2	9.0	10.9	suitable
3	Schwaben	loamy sand	0.3	0.7	1.3	unsuitable

As can be seen from the table, the number 3 soil is unsuitable as its nitrification rate is so low. The first soil can transform the added ammonium within 14 days, and the second within 28 days.

A soil such as the third should, thus, not be chosen for nitrogen cycle studies without serious thought as to its nitrogen transformation capacity.

Is alfalfa green meal suitable as an organic N-substrate?

In the Recommendations published earlier (Greaves et al., 1980), it is recommended that the release of ammonium could be studied in the same way as described

for the mineralization of carbon, that is following the
addition of 0.5% alfalfa green meal. To investigate
whether this substrate is suitable we performed several
trials in our laboratories. So far, the results have
shown that the amount of mineralized nitrogen due to the
addition of 16 mg N/100 g soil did not increase during a
28 days trial period (Figure 3).

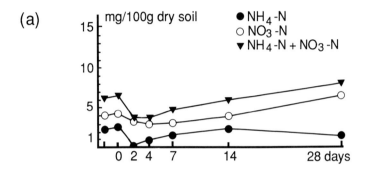

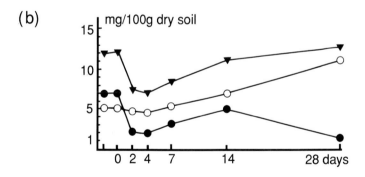

Figure 3   Mineralization of alfalfa green meal

        a) in a fresh loam soil
        b) in a German standard soil (2.2)

On the other hand it has been observed with both
soils, the fresh Hattersheim soil and the LUFA standard
soil 2.2, that the nitrogen concentration (the sum of

ammonium-N and nitrate-N is very often indicated as N_{min} which stands for total mineralized nitrogen at the start of the trial is almost as high as after 28 days.

A ring trial brought similar results (Table 3):

Table 3   Mineralization of alfalfa green mean: Results of a ring test (16 mg N/100 g soil)

Laboratory	Soil	mg NH4-N + NO3-N/100 g soil days after starting the experiment			
		0	7	14	28
Bayer	Laacher Feld 117	6.1	4.3	5.2	6.5
Celamerck	Ingelheimer Sand	0.0	0.0	0.0	0.7
Hoechst	Standard Soil 2.2	12.2	8.8	11.5	12.6
BASF	Neuhofen loamy sand	0.0	0.0	0.1	2.6

The reason why alfalfa is not suitable as an organic nitrogen substrate is not understood at present. The C/N ratio of this substrate is normally <20, in our sample about 13, and of the soluble parts 1.8 (Table 4).

Table 4   C/N-analysis of alfalfa green meal

	C %	N %	C/N ratio
dry material	39.9	3.2	12.5
soluble parts (buffer solution, pH 8)	5.9	3.2	1.8

A possible explanation is that the bioavailability of the extractable parts is so good that the microbial population increases and, therefore, nitrogen is assimilated during the chosen time of the experiment (28 days). Since alfalfa has been found to be unsuitable we searched for new substrates in further trials.

Is horn meal suitable as an organic N-substrate?

In a ring trial it has been shown by all participants that the major part of the 20 mg nitrogen in the 166 mg horn meal which was added to 100g soil was mineralized within 28 days (Figure 4). Observations over a longer period of, for instance, 56 days brought no change in the results.

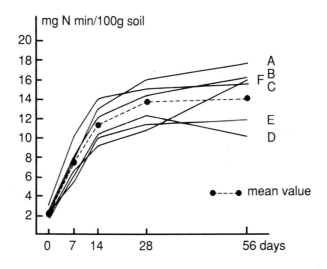

Figure 4   Ammonification of horn meal :
            Results of a ring trial
            (A - D, participants)

In the graph the mineralized amounts of nitrogen (N_{min}) are plotted. Results deviating from the mean can not be explained by methodological difficulties but by varying temperatures during the trials.

To illustrate the course of the nitrogen mineralization one example was chosen, in the following table 5.

Table 5   Mineralization of horn meal
           (20 mg N/100 g soil)

Treatment		mg N/100 g soil days after starting				% N
	– N	0	7	14	28	min*
Control	NH4–	1.13	0.10	0.09	<0.05	69
	NO3–	0.96	9.80	13.10	15.87	
5 x application rate of a pesticide	NH4–	1.08	0.67	0.09	<0.05	71
	NO3–	0.98	9.67	13.59	16.20	
1 ppm Nitrapyrin	NH4–	1.03	9.64	11.67	10.10	64
	NO3–	1.00	1.22	2.04	4.78	

$$N\ min* = \frac{(N\ min_t - N\ min_{to}) \cdot 100}{20\ mg\ horn\ meal - N}$$

In all soils, independant of nitrogen treatment, about 70% of the added organic nitrogen was mineralized within one month. These results show that in contrast to alfalfa green meal, horn meal is a very good organic nitrogen substrate for ammonification and nitrification investigations. This presents a further advantage of this substrate, in that both transformation processes can be studied in a single trial. Furthermore, calculation from the data obtained is very easy. In Table 6 the formulae for determining ammonification and nitrification as well as the percentage transformation in comparison to the control are given.

Table 6   Calculation of the ammonification and nitrification
           of horn meal N
           (Raw data from Table 4)

Treatment	mg N/day	% of control	mgNO3+ N/day	% of control
Control	0.49	100	0.53	100
5 x application rate of a pesticide	0.51	104	0.54	102
1 ppm Nitrapyrin	0.46	94	0.14	26

formulation
for calculation

$$AR = \frac{N \min_t - N \min_{to}}{t} \qquad NR = \frac{NO_3-N_t - NO_3-N_{to}}{t}$$

AR = ammonification rate   NR = nitrification rate

Using the ammonification and nitrification rate
(absolute or percentage) it can be ascertained whether or
not a pesticide affects these nitrogen transformations.
The chosen example demonstrates, that even with a
fivefold application rate, no negative influence can be
seen. This example should be considered as taken at
random from many trials in which different pesticides had
been applied and which have all behaved similarly.

By following this easy evaluation it is possible to
assess the side effect of pesticides on the nitrogen
transformation with the help of the scheme developed by
Domsch and co-workers.

If a pesticide influences nitrification negatively
after a period of 1 month, the duration of the trial
should be prolonged as long as the nitrate contents in
the control and the treated soil are identical. Under
these circumstances it should be considered whether an
additional application area should be found for this
compound. From the ecological point of view this
influence on the nitrification is all the more important
because the upper permissible limit for nitrate in
European drinking water is 50 mg/l and, in that way, a

potential burden by nitrate could be prevented.

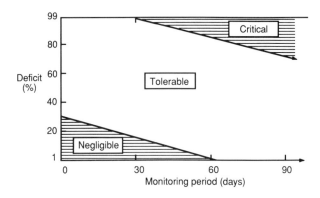

Figure 5   Assessment of side effects of pesticides on the
            soil microflora
            (From Domsch et al., 1983)

Field trials would only be meaningful if a non-tolerable effect according to the Domsch-scheme, on nitrogen transformation were detected, after a normal application rate in the model trial had been carried out.

Generally speaking, field trials are not very useful, since they are not easily reproduced due to varying climatic influences. Furthermore, the climatic stress on the soil microorganisms is usually far more severe than that of an agrochemical. Finally to assess a possible side-effect of a pesticide on the carbon or nitrogen-mineralization the experimental conditions must be identical in all trials to enable a proper comparison of results, obtained with different pesticides.

CONCLUSIONS

1. German Standard Soils are unsuitable for nitrogen transformation trials without prior biological examination.

2. A soil is suitable if it can transform 10 mg N/100 g soil within 2 - 4 weeks. This must be clarified in preliminary trials.

3. Alfalfa green meal is unsuitable as an organic nitrogen substrate, since the small mineralized amounts of nitrogen are obviously assimilated by microorganisms.

4. Horn-meal is an appropriate nitrogen-substrate, because within 4 weeks about 70 % is mineralized. A further advantage of this substrate is that it can be used to determine both ammonification and nitrification in a single trial.

REFERENCES

Domsch, K.H., Jagnow, G. and Anderson, T.H., 1983, Residue Reviews, 86, 66.

Greaves, M.P., Poole, N.J., Domsch, K.H., Jagnow, G. and Verstraete, W., 1980, Recommended Tests for Assessing the Side-Effects of Pesticides on the Soil Microflora. Technical Report, Agricultural Research Council, Weed Research Organisation, No. 59.

CHAPTER 9

AMMONIFICATION AND NITRIFICATION.
LABORATORY VERSUS FIELD

H. van Dijk

INTRODUCTION

In the report of the 1979 Workshop on Side-effects of Pesticides on the Soil Microflora (Greaves et al., 1980) it is concluded that 'ammonification and nitrification', as part of the nitrogen-mineralization process, are useful criteria to study potential side-effects of pesticides. Therefore, laboratory tests with regard to these processes were recommended as a requirement for regulatory assessment in The Netherlands, assuming that these tests would serve as 'yardsticks' for the impact of pesticides applied in the field.

The questions at issue now are whether this assumption has been justified by experience and whether and how laboratory data can be translated to field conditions. Unfortunately, reports of laboratory measurements of ammonification and nitrification in soils treated and untreated with pesticides, accompanied by simultaneous measurements in the field with the same soil are very scarce. Mostly laboratory tests and field experiments have not been carried out with the same soil at the same time; in such cases the different results may partly be caused by differences in soil properties and/or populations. I am not aware of field experiments with crops where effects of pesticides on both processes were deduced from a complete nitrogen balance sheet, taking into account nitrogen uptake by the crop, nitrogen leaching, denitrification, nitrogen fixation, in short the complete nitrogen cycle in the soil. If such data were available, the conclusion is based on our experience with soil fumigation in autumn where the situation is indeed less complicated because no crop is present. Field conditions usually differ markedly from those in the laboratory experiments (Table 1).

Table 1   Difference between experimental conditions in the laboratory and in the field

	Laboratory	Field
Pesticide distribution	even	uneven
Temperature	constant (20°C)	varying daily seasonal
Moisture	constant (pF 2.5)	varying short-term seasonal
Aeration	sufficient	varying
Pesticide dissipation rate	rel. high rate of transform. (20°C)	lower rate of transform., but other ways of dissipation possible (volatilization, leaching)
Presence of plants	no plants	usually cropped

FIELD VS LABORATORY EXPERIMENTS

Distribution of pesticides in soil

The distribution of pesticides in the soil in field situations is always uneven, unless extremely persistent pesticides gradually accumulate after repeated applications. In laboratory tests a homogeneous distribution is ensured. Actual distribution patterns of fumigants in the field, for example, show that the concentration in the surface layer is always practically zero. Surface-applied herbicides may not reach a significant concentration below 10 or 20 cm depth. Thus, part of the biologically active soil remains 'untreated', which is of great importance for the recovery potential of natural transformation processes. In those cases the actual rate of recovery is often increased by soil mixing (tillage).

Temperature

The (differences in) temperature dependence of the rates of ammonification and nitrification are known, so extrapolation from laboratory to field conditions is, in principle, possible. The temperature dependence of the dissipation rate of the pesticide, however, is a complicating factor. A prolonged inhibition of nitrification after soil fumigation in autumn (Lebbink and Kolenbrander, 1974) may only be due to a low soil temperature or partly also due to a retarded dissipation of the fumigant at that temperature.

Moisture content and aeration

Moisture content and aeration of the soil in field situations also vary. Within the range of moisture contents that normally occur in the field, the rate of ammonification may vary by a factor of 3 to 5. Ammonification does not stop under anaerobic conditions. Little or no nitrification takes place in dry or extremely wet soils (periods of extreme drought or excessive rainfall are, however, rare and normally not long in our region). In laboratory tests the occurence of anaerobic sites leading to denitrification is avoided. Field data on nitrate contents may not be 'clean' in this respect. The moisture content usually also influences the dissipation rate of the pesticides and often their distribution pattern. The presence of anaerobic sites may seriously affect their transformation rate and, thus, the recovery rate of nitrogen transformations.

Other pathways of pesticide dissipation

Other possible pathways of pesticide dissipation in the field (leaching, volatilization) also introduce uncertainty with regard to the possibility of translating laboratory data on nitrogen transformations, not only because of differences in actual disappearance rate but also because of the consequences for distribution and, thus, recovery potential.

Presence of plants

The presence of plants and the type of plants present are modifying factors for the soil population. The presence of plants (rhizosphere) affects the metabolic processes in the soil, presumably also the transformation rate of pesticides and, thus, their possible side-effects.

DISCUSSION

The adequacy of the recommended laboratory tests was checked by Van Faassen and Lebbink (1984) by comparing the effect of soil fumigation with high rates of 1,3-dichloropropene (600 l/ha) or chloroform (1200 l/ha) in the field with that of chloroform treatment in the laboratory. The relevant results are summarized in Table 2. Size and duration of the effect (if present) of fumigation on ammonification and nitrification, as observed in incubation measurements with soils treated in the field or in the laboratory, differed considerably. A negative effect on ammonification was only observed in tests with soils fumigated with chloroform in the laboratory and amended with lucerne meal. In the field the recovery apparently was quick, probably because there,

Table 2   Effect of soil fumigation in the field and in the laboratory on ammonification and nitrification

	Field fumigation with dichloropropene/ chloroform at high rates		Fumigation with chloroform in the laboratory	
	sandy soil	sandy loam soil	sandy soil	sandy loam soil
Ammonification of soil org. nitrogen	n.a.	+	n.a.	+
Ammonification in lucerne amended soil	+	+	-	-
Nitrification of (NH4)2SO4	-/n.a.	s	-	-

+   = activity increased
n.a. = not affected
s   = limited effects and/or rapid recovery
-    = strong effects with slow recovery

soil life is never completely exterminated; the reasons are an uneven distribution of fumigant, and/or unaffected species rapidly taking over from eliminated ones, or (particularly with short-lived pesticides) ammonification rapidly being restored from less vulnerable rest-structures of the hit species. Judging from this experience, the laboratory tests on ammonification in fact still provide no information about the effect of pesticides on ammonification in the field.

Apart from that, the maximum effect of pesticides, when applied at recommended rates, on ammonification in the field (a nitrogen flush in the order of 10 kg nitrogen/ha) is agronomically rather insignificant.

Contrary to ammonification, the effect of soil fumigation (and of pesticide use in general) on nitrification in laboratory tests and in the field qualitatively is always the same and, if occurring, it may be of agronomic significance. For example, the amount of nitrogen saved from leaching in wet winters may be as high as 50 kg/ha on sandy soils after autumn fumigation with dichloropropene (Lebbink and Kolenbrander, 1974) and even more when animal manure has been applied.

Concerning duration of nitrification inhibition in the field two factors are of prime importance, viz., the actual persistence of the inhibitor and the time of the year in which it is applied.

CONCLUSIONS

The recommended ammonification tests are inadequate for assessing the effect of pesticides on this process, or on the soil microflora generally, in the field.

The nitrification test should be maintained because of its sensitivity and the agronomic significance of this process.

For the time being there are few prospects for predicting the field effect from laboratory data; there are too many uncertainties.

A point of further discussion is whether laboratory tests on nitrification should be laid down for all pesticides, irrespective of their usage pattern (dosage rate, mode and time of application), mode of inhibitory action and actual persistence in the upper soil horizon.

REFERENCES

Greaves, M.P., Poole, N.J., Domsch, K.H., Jagnow, G. and

Verstraete, W., 1980, Recommended Tests for Assessing the Side-effects of Pesticides on the Soil Microflora, Technical Report, Agricultural Research Council, Weed Research Organisation, No. 59.

Lebbink, G. and Kolenbrander, G.J., 1974, Agriculture and Environment 1, 283.

Van Faassen, H.G. and Lebbink, G., 1984, Plant and Soil 76, 389.

CHAPTER 10

SYMBIOTIC NITROGEN FIXATION

S. Horemans, K. De Coninck, R. Dressen and K. Vlassak

INTRODUCTION

This text has been written to introduce a discussion on both the need for and the relevance of testing side effects of pesticides on nitrogen fixation. In consequence the formulation of the statements is rather straightforward and the selection of the literature, by no way complete, is unavoidably biased towards the authors point of view. We gratefully made use of a number of reviews on this topic (Anderson, 1978; Greaves et al., 1978; Greaves et al., 1980a, 1980b; Hohnen, 1978; Ottow, 1985).

ACTUAL AND POTENTIAL IMPORTANCE OF SYMBIOTICALLY FIXED NITROGEN

Burns and Hardy (1975) estimated an annual input of 89 x 10^6 metric tons/year of biologically fixed nitrogen under crops and permanent meadows. This is about double the annual amount of industrially produced nitrogen. Most of the biological fixation can be ascribed to Rhizobium spp., bacteria living symbiotically with legumes. The discussion will be devoted to this group of crops, which account for only 9% of the combined world dry matter yield of cereals and legumes, but which constitute as much as 24% of the total protein yield (Sigurbjörnson, 1984). La Rue and Patterson (1981) summarized some valid estimates of nitrogen fixed by forage legumes, soybeans and pulses. These estimates range from very low up to over 250 kg nitrogen/ha. year and are in good accordance with the average figure of 140 kg/ha year reported by others (Burns and Hardy, 1975;

Fried and Broeshart, 1975; Döbereiner, 1978).

However, extrapolation of these figures, obtained in experimental plants, to actual yields under farming conditions is not proper, especially in view of the variability of environmental factors such as soil conditions, moisture, temperature and improper use of fertilizers and herbicides, which may depress nitrogen-fixing capacity. Burris (1978) stated that half this amount would be a more realistic figure and La Rue and Patterson concluded "there is no single legume crop for which we have valid estimates of the nitrogen fixed in agriculture". Nevertheless, the data referred to give some idea of the potential contribution which can be made.

Our knowledge on the actual impact of nitrogen fixation in tropical regions is even more limited. Vincent (1982) gave an overview of common legumes in Asian and Pacific regions together with recordings of nodulation (as a parameter of the presence of Rhizobium) and effectivity. There is good hope that as a result of international projects (e.g. the USAID Niftal project in Hawaii, the FAO/IAEA project on mutation breeding for improved nitrogen fixation in grain legumes, other microbiological research centres supervised by the UNEP/UNESCO/CRO panel) more accurate data and reports of substantial yield improvements will soon be available. Although, in these countries, the barriers to maximal utilisation are numerous, the organisations mentioned (as well as many others) believe that Rhizobium will, in the very near future contribute substantially to the fulfilment of the urgent need for proteins. It has been argued that symbiotic nitrogen fixation will play also a major role in reforestation projects both in temperate zones (Gordon, 1984) and in tropical regions (Dommergues et al., 1984).

In addition to an input of nitrogen to the soil, biological nitrogen fixation offers other advantages. The absence of direct hazards to the environment, which may be the case with excessive application of chemical fertilizers, considerably reduces the real cost paid by the community for protein production. In the same context, it is commonly accepted that the use of mineral fertilizers and herbicides are interdependant inputs. Thus greater use of the former requires increasing use of the latter. Symbiotically fixed nitrogen is directly transferred to the plant of interest and is not available for growth of unwanted weeds.

Finally we are only beginning to discover factors leading to successful establishment of specific strains of Rhizobium. The worldwide pool of presently undiscovered or neglected adaptations of this particular bacteria can be a treasure for mankind in the future.

The use of pesticides has become an accepted, economically essential and, at the present state of art, unavoidable practice in modern agriculture. There is time for legume crops, mixed legume/non-legume crops and for cropping systems where legumes are a normal constituent in the rotation. Whereas negative interactions of these plant protection chemicals with biological nitrogen fixation should be avoided, the availability of safe products or products stimulating an effective symbiosis can be of great economical value for farmer, society and manufacturers alike.

Whatever the policy of those responsible for application, production or registration of these protection chemicals, they need a reliable, comparable and relevant test system. The development of, and agreement on such a test system is be a priority challenge for all scientists involved.

CHOICE OF AN APPROPRIATE PARAMETER FOR ESTIMATING EFFICACY OF THE PLANT-RHIZOBIUM SYMBIOSIS.

The establishment and effective legume/Rhizobium symbiosis is not a single process but is the result of a series of mutually controlled interactions including infection, nodule initiation, bacteroid development, development of the nutrient flow between endosymbiont and host, persistence and ultimate senescence of nodule functions. Pattern specification proceeds in a strictly time dependent way and the overall process must be accomplished within a plant's lifetime. Adverse environmental conditions or in casu pesticides can interfere with the delicate equilibrium between the host and the potential phytopathogenic bacteria (Verma and Nadler, 1984). Their impact tends to be either negligible, when homeostasis can be maintained by the system, or delaterious when the same mechanisms lose their control or deliberately abort the process. For this reason, the combined effects of naturally occuring stresses and an additional man-made stress, or of two interacting man-made stresses are largely unpredictable.

Since effective biological nitrogen fixation is dependent on successful achievement of all the preeding steps we cannot use any one of these physiological, biochemical or morphological functions in isolation as a valuable indication of overall efficacy.

A few literature reports will clarify the statements made above. Plate counts (Curley and Burton, 1975) showed that R. japonicum retained the ability to multiply in contact with captan or carboxim. In contrast, nodule counts at 2 weeks age demonstrated that both pesticides were toxic. In a pot experiment (Mallik and Tesfai, 1984) the insecticides acephate, diazinon and toxaphene did not affect growth or nodulation of soybeans. However, all three products decreased the nitrogen content of the shoots. Rennie et al. (1985), tested the effect of captan-based, seed applied pesticides on 3 different crops (peas, lentils, faba beans). They scored nodulation ratings, acetylene reduction activity as well as plant yield. They concluded that "Nodule numbers may or may not correlate with actual nitrogen fixed depending on cultivar, plant growth stage, temperature regime, etc. and thus are poor indicators of nitrogen fixation".

Another illustration comes from a study on the effect of soil acidity on symbiotic nitrogen fixation in clover (Thorton and Davey, 1983). A low pH interfered with the timing of nodulation, the latter being, indeed, a transient property of root cells (Rolfe and Shine, 1984). Consequently, nodules were predominantly found on the lateral roots, instead of on the crown roots. To cite the authors, the latter nodules "tend to be more pathogenic than symbiotic" and overall nitrogen yield turned out to be reduced.

Some authors do use dry matter yield as a parameter to estimate the success of inoculation. However, Hardarson et al. (1984), in a study on soybeans, found that total dry matter yield, although sometimes correlated to effectiveness in nitrogen fixation, especially where nitrogen is the only limiting factor for growth, was not likely to be sensitive enough to measure benefits from inoculation.

In view of the endogenous control mechanisms and taking into account the time limits imposed by crop assessing, the use of an ecological "yardstick" for comparing the effects of applied pesticides does not make sense with respect to effects on symbiotic nitrogen fixation. The latter approach (Domsch, in Greaves et al., 1980) seems to be restricted to free-living organisms. The most straightforward interpretation of overall, integrated efficacy must be obtained by using an economical "yardstick" i.e. an estimate of the total amount of nitrogen fixed symbiotically and transferred to the plant at the end of the growth period of interest. How this can be done will be discussed later.

CHOICE OF THE CULTURE CONDITIONS

There is general agreement that pesticide side-effect testing should be performed in soil. Indeed, the activity and relative abundance of both the Rhizobium strain of interest and of other micro-organisms competing for the same niches are completely disturbed in artificial media. Also, the fate of the pesticides applied (concentration, distribution, metabolism, masking or activation of metabolically active groups) is altered and can even be different for different artificial media (Jagnow et al., 1979). This can be illustrated by data from Smith et al. (1978), who tested some organophosphate and carbamate insecticides. Significant differences were observed in toxicity which were dependent on the growth substrate used. More specifically the approximate field-rate applications of carbofuran, carboryl and aldicarb tended to depress nitrogenase activity in soil (up to 63% of normal activity) but not in vermiculite (90% of normal activity). Finally, root morphology and developmental pattern as well the physiological activity (water economy, photosynthetic capacity, sink-source relationships) of the host plant are altered. It can be argued that, "conventional laboratory testing of side effects is an expensive, illogical and unnecessary exercise" (Greaves and Wingfield, 1983).

The use of pot experiments has been advocated (Greaves et al., 1980). However soil must be removed from the field, transported, stored and prepared for use. Test plants, a necessary prerequisite for testing the symbiotic process, grow under artificial light, temperature and humidity conditions. For these reasons pot experiments can be criticised on the same ground as the use of any other artificial medium. In pot experiments, Dunigan et al., (1972), reported an inhibition of nodulation by several herbicides. In a three year field study these same herbicides did not reveal any detrimental effects.

We are forced to conclude that, if the goal is to provide an estimation of actual interactions in the field, tests should be performed under representative field conditions. One additional argument will be developed in the next section.

CHOICE OF THE BACTERIAL PARTNER

Rhizobium species are defined on the basis of their host range, particular legumes being nodulated only by specific Rhizobium strains. Thus, R.trifolii nodulates clovers, R.meliloti lucerne and R.japonicum soybeans. In contrast to the narrow specificity shown by these species on predominantly temperate legumes, another large group, known as the "cowpea miscellany" (the Brady-Rhizobium group), covers those Rhizobia which have a broad range and can usually infect a variety of tropical legumes. This distinction is parallelled by differences in growth rate under laboratory conditions, type of nodules formed and export products of the bacteroids to the host plant (Rolfe and Shine, 1984). For each subspecies a large range of naturally found or man-selected strains are present. Remarkable differences in sensitivity to pesticides occur between subspecies as well as between different strains of the same group (Heinnanen-Tanski et al., 1982). Consequntly the choice of the bacterial partner can influence the final evaluation of a specific product. The only valid parameter we found for selecting a test organism is its actual importance. Any soil can contain a mixture of strains differing in their susceptibility for pesticide action and in their capacity to have an effective symbiosis with the crop of interest (e.g. the indigenous and the inoculated population). Selective interference with only one of them will alter final yield. More clearly stated the species on which a particular pesticide should be checked must be tested under representative environmental (i.e. agricultural) conditions.

CHOICE OF THE INOCULATION METHOD

In most cases, nodulating bacteria are present in the soil. Nevertheless alien strains can have greater potential value. In these cases inoculation is required. Since the indigenous population is often superior with respect to survival and competitiveness (Trinick, 1982; Dunigan et al., 1984), problems involving introduction and stable establishment of Rhizobium spp. are numerous. It is achieved either by applying inoculum to the seed before sowing or directly into the seed bed. Since pesticide activity is a matter of both contact and concentration these methods represent two completely different situations with respect to seed dressings of plant protection chemicals agents. Application of inoculum directly into the seedbed is more readily comparable to situations where Rhizobium is present in the soil. In our opinion, the complications of strain introduction during experimentation increase costs as well as possible errors (e.g. Duczek and Buchan, 1981) and should be avoided in pesticide screening. Wherever possible, the Rhizobium of interest must be firmly established in the soil. If not, seed bed inoculation should be preferred. This view is strengthened by recent increased interest, partly due to the need for chemical seed protection, in seed bed inoculation using dispensed and granular inoculants, the latter giving excellent results. For example, with Phaseolus vulgaris, and despite a low soil pH, granular inoculants induced nodulation which was superior to that of seed applied inoculants (Graham et al., 1980). The authors recommended this method to circumvent negative effects they found when thiram or PNCB was used as a seed dressing agent in combination with seed inoculation. Rennie and Dubetz (1984) concluded that the seed applied fungicides thiram, captan, and carbathin could be used safely for soybeans with granular Rhizobium inoculation, whereas they depress nitrogen fixation with seed-applied inocula. Also, Chamber and Montes (1982) advised application of the inoculum into the seed bed, instead of to the seed, when captan-treated soybeans had to grow under increased environmental stress conditions. Dunigan et al. (1984) used soil inoculations. This method allowed the introduction and stable establishment of a non-indigenous strain of R.japonicum. In practice we suggest that compatibility with seed inoculation should not be made obligatory. Ways to circumvent this technique do exist. However, since it is a commonly used technique, a feasible registration policy could be the introduction of precautionary warning : "When seed applied, this product is incompatible with the survival of seed inoculated Rhizobia". This text can be omitted when the company or an independent organization has demonstrated the opposite. Procedures for testing the survival can be found in the literature (e.g. Curley and Burton, 1975; Graham et al., 1980; Chamber and Montes, 1982).

METHODOLOGICAL ASPECTS AND EXPERIMENTAL DESIGN

From the data collected in the field experiment the experimentor must be able to decide:

- Whether the biological fixation process was active in the field experiment done;
- Whether any negative or positive effect measured was due to an interaction with biological nitrogen

139

fixation.

Therefore, the following data are needed:

- Nitrogen content of a non-fixing control plant both at low and high nitrogen levels (e.g. 10 and 150 kg/ha, properly fractionated).
- Nitrogen content of the untreated legume at both high and low levels of nitrogen. The plant protection chemical should be replaced, if possible, by another appropriate control method proven to be compatible with nitrogen fixation.
- Nitrogen content of the pesticide treated legume at both high and low nitrogen-dose.

The number of trials carried out must be sufficient to give a representative sample of the crop concerned. Trials should cover the main geographical region in which the crop is grown.

Three methods for obtaining integrated values of nitrogen fixation by field - grown legumes are in use; "A"-value, isotope dilution and the comparison of total nitrogen in "fixing" and comparable "non-fixing" plants (the difference method). All 3 methods require the inclusion of a non-fixing control to estimate the relative contribution of soil and fertilizer nitrogen. Both the isotope dilution and "A"-value technique (Fried and Broeshart, 1975; Legg and Slogger, 1975) require labelling of the soil nitrogen pool with 15N-enriched fertilizer. These methods are undoubtedly the best available (Rennie, 1982) but require expensive equipment and highly trained personnel. The difference method requires only nitrogen determinations which can be done by standard Kjeldahl determination. Reliability has been tested several times (Vasilas and Ham, 1984; Hardarson et al., 1984) and excellent agreement was found with isotopic estimations. The "difference method" seems, therefore, to be an acceptable choice.

Several possible non-fixing control plants have been proposed (Bell and Nutmann, 1971); ineffective nodulating (I) lines, uninoculated nodulating lines (O), non-nodulating isolines (N), and non-legumes (G). Both (I) and (O) are not useful as we have advocated the use of fields containing effective soil-borne Rhizobia. Nevertheless, they have been shown to be the best non-fixing controls available (Rennie, 1982). The use of (N) was, until very recently, restricted to soybeans but, partly under the influence of the FAO/IAEA group on

140

mutation breeding for improved nitrogen fixation, they have now become available for some other species such as pea (Kneen and La Rue, 1984) and alfalfa (Vance et al., 1984). Further confirmation of their validity is required and for the present time at least, non-legumes should be used as the alternative. Hordeum vulgare (Fried and Broeshart, 1975) or Lolium perenne (Domenach et al., 1979) have been used in temperate climate. Some authors (Rennie, 1982) rely even more on non-legumes than on (N) as a control for absolute quantifications. Anyway appropriate control plants for use in tropical and subtropical cultures are needed urgently. They should have a similar seasonal pattern of soil nitrogen uptake and a maturation date comparable to the legume crops of economical importance.

CONCLUSIONS

(1) The only valid parameter for estimating the efficacy of the overall process is the parameter of interest i.e. the total amount of nitrogen fixed and transferred to the plant. This should be tested under representative agricultural conditions using strains firmly established in the soil.

(2) If inoculation is necessary, seed bed inoculation is preferable to seed inoculation when pesticide seed treatments are used.

(3) If the appropriate control plants can be found and experimental organisation is respected the "difference" method seems to be of sufficient accuracy, sensitivity and simplicity to be the method of choice.

ACKNOWLEDGEMENTS

The authors wish to thank Prof. R.H. Clements, Dept. of Crop Science, University of Peradeniya, for his interest and suggestions in the present study; and Mrs. J. Pierre and M.J. Struyven for the secretarial work.

REFERENCES

Anderson, J.R., 1978, in Pesticide Microbiology, edited by I.R. Hill and S.J.L. Wright (London, Academic Press) p. 313.

Bell, F. and Nutman, P.S., 1971, in **Nitrogen Fixation in Natural and Agricultural Habitants**, edited by T.A. Lie and E.G. Mulder (Plant and Soil, Special Volume), p. 231.

Burns, R.C. and Hardy, R.W.F., 1975, in **Nitrogen Fixation in Bacteria and Higher Plants**, edited by R.C. Burns and R.W.F. Hardy (Berlin, Springer-Verlag).

Burris, R.H., 1978, in **Environmental Role of Nitrogen-Fixing Blue-Green Algae and Symbiotic Bacteria**, edited by U. Granhall (Stockholm, NFR Editorial Service).

Chamber, M.A. and Montes, F.J., 1982, **Plant and Soil**, 66, 353.

Curley, R.L. and Burton, J.C., 1975, **Agronomy Journal**, 67, 807.

Döbereiner J., in **Limitations and Potentials for Biological Nitrogen Fixation in the Tropics**, edited by J. Döbereiner, H. Burns and A. Hollaender, (New York Plenum Press).

Domenach, A.M., Chalamet, A., and Pachiaudi, C., 1979, **Comptes Rendus Hebdomadaires des Seances de L'Academie des Sciences**, 289, 291.

Dommergues, Y.R., Diem, H.G., Gauthier, D.L., Dreyfus, B.L. and Cornet, F., 1984, in **Advances in Nitrogen Fixation Research**, edited by C. Veeger and W.E. Newton (Den Haag, Martinus Nijhoff/Dr. W. Junk Publ.), p. 7.

Duczek, L.J., and Buchan, J.A., 1981, **Canadian Journal of Plant Science.**, 61, 727.

Dunigan, E.P., Bollich, P.K., Hutchinson, R.L., Hicks, P.M., Zaunbrecher, F.C., Scotts, G. and Mowers, R.P., 1984, **Agronomy Journal**, 76, 463.

Dunigan, P.E., Frey, P.J., Allen, L.D. and McMahon, A., 1972, **Agronomy Journal**, 64, 806.

Fried, M. and Boeshart, H., 1975, **Plant and Soil**, 1975, 43, 707.

Fried, M. and Middleboe, V., 1977, **Plant and Soil**, 47, 713.

Graham, P.M., Ocampo, G., Ruiz, L.D. and Duque, A., 1980, Agronomy Journal, 72, 625.

Gordon, J.C., 1984, in Nitrogen Fixation Research, edited by C. Veeger and W.E. Newton (Den Haag, Martinus Nijhoff/Dr. W. Junk Publ.) p. 15.

Greaves, M.P., Lockhart, L.A. and Richardson, N.G., 1978, Proceedings 1978 British Crop Protection Conference - Weeds, p. 581.

Greaves, M.P., and Malkomes, H.P., 1980, in Interactions Between Herbicides and the Soil, edited by R.J. Hance (London Academic Press) p. 223.

Greaves, M.P., Poole, N.J., Domsch, K.H., Jagnow, G. and Verstraete, W., 1980, Recommended Tests for Assessing the Side-Effects of Pesticides on the Soil Microflora, Technical Report, Agricultural Research Council, Weed Research Organisation, No. 59.

Greaves, M.P., and Wingfield, G.I., 1983, Pesticide Science, 14, 634.

Hardarson, G., Zapata, F. and Danso, S.K.A., 1984, Plant and Soil, 82, 369.

Heinonen-Tanski, H., H., Oros, G. and Kecskes, M., 1982, Acta Agricultura Scandinavia, 32, 283.

Jagnow, G., Heinemeyer, O. and Draeger, S., 1979, Ecotoxicology and Environmental Safety, 3, 152.

Jansen van Rensbur, H. and Strijdom, B.W., 1984, South African Journal of Plant and Soil tijdskr. Plant Grond, 1, 135.

Johnen, B.G., Drew, E.A. and Castle, D.L., in Soil-Borne Plant Pathogens, edited by B. Schippers and W. Can (London, Academic Press) p. 513.

Kneen, B.E. and La Rue, T., 1984, in Advances in Nitrogen Fixation Research, edited by C. Veeger and W.E. Newton (Den Haag, Martinus Nijhoff/Dr. W. Junk Publ.), p. 599.

La Rue, T.A. and Patterson, T.G., 1981, Advances in Agronomy, 34, 15.

Legg, J.O. and Sloger, C., 1975, Proceedings International Conference Stable Isotopes, p. 661.

Mallik, M.A.B. and Tesfai, K., 1985, Plant and Soil, 85, 33.

Ottow, J.C.G., 1985, Naturwissenschaftliche Rundschau, 38, 181.

Rennie, R.J., 1982, Canadian Journal of Botany, 60, 856.

Rennie, R.J. and Dubetz, S., 1984, Agronomy Journal, 76, 451.

Rennie, R.J., Howard, R.J., Swanson, T.A. and Flores, G.H., 1985, Canadian Journal of Plant Science, 65, 23.

Rolfe, B.G. and Shine, J., in Genes Involved in Microbe Plant Interactions, edited by D.P.S. Verne and T.H. Hohn, (Wien, Springer-Verlag) p. 44.

Sigurbjörnsson, B., 1984 in Breeding Legumes For Enhanced Symbiotic Nitrogen Fixation, edited by G. Hardarson and T.A. Lie (Plant and Soil, Special Volume), p. 3.

Thornton, F.C. and Davey, C.B., 1983, Agronomy Journal, 75, 557.

Thurlow, D.L. and Hiltbold, A.E., 1985, Agronomy Journal, 77, 432.

Trinick, 1982, in Nitrogen Fixation in Legumes, edited by J.M. Vincent (Australia, Academic Press) p. 229.

Vance, E.P., Heidsel, G.H., Barnes, D.K., 1984, in Advances in Nitrogen Fixation Research, edited by C. Veeger and W.E. Newton (Den Haag, Martinus Nijhoff/Dr. W. Junk Publ.), p. 565.

Vasilas, B.L. and Ham, G.E., 1984, Agronomy Journal, 76, 759.

Verm, D.P.S. and Nadler, K., 1984, in Genes Involved in Microbe-plant Interactions, edited by D.P.S. Verma and T.H. Hohn (Wien, Springer Verlag) p.58.

Vincent, J.M., 1982, in Nitrogen Fixation in Legume,

edited by J.M. Vincent (Australia, Academic Press) p. 263.

CHAPTER 11

ESTIMATION OF MICROBIAL BIOMASS
BY STIMULATION OF RESPIRATION

H. Van de Werf & W. Verstraete

INTRODUCTION

Current methods to estimate microbial biomass in soils are most frequently based on soil fumigation (Jenkinson and Powlson, 1976), respiratory response (Anderson and Domsch, 1978), ATP measurement (Paul and Johnson, 1977; Verstraete et al., 1983) and microcalorimetry (Sparling, 1981). These methods have been used for different purposes: to study the decomposition of organic matter (Ladd et al., 1981; Marumoto et al., 1982; Voroney and Paul, 1984), the accumulation of nutrients in biomass (Anderson and Domsch, 1980), the influence of storage on the soil microbial biomass (Anderson et al., 1981; Doelman and Haanstra, 1979; Haanstra and Doelman, 1984). All these methods are generally calibrated on the basis of the fumigation-reinoculation principle. However, the latter biomass values are approximations of unknown true values and furthermore comprise both active and surviving cells.

In this chapter, the changes in the biokinetic parameters of the active soil biomass, as a result of the impact of a number of typical soil pollutants are described. The method is based on monitoring, in a continuous respirometer, the oxygen uptake of soil samples supplied with readily metabolisable organic substrates. These curves, which are subsequently analysed, conform to convential microbial growth kinetics (Van de Werf and Verstraete, 1986a,b). This permits the derivation of estimates, not only of the quantity of active soil microbial biomass, but also of the maximum specific growth rate, the substrate affinity, the maximum cell yield and the cell maintenance coefficient of that biomass.

MATERIALS AND METHODS

Respiration experiments were done with three different soil types, namely Gistel, Bierbeek and Bonheiden soil. Table 1 gives some physico-chemical properties of these soils.

The soils were collected from the upper 30 cm of field plots, 'homogenized' by sieving through a 4 mm screen and adjusted to 75% field moisture capacity. The soils were incubated in the dark in plastic bags covered with polyethylene films at 20°C. After 14 days of equilibration, the soils were subjected to disturbance as follows:

Increase of soil acidity

The acidity of the soils was increased by one pH-unit. This was done after experimental determination of the soil pH-curves. To decrease the soil pH, the soils were amended with diluted H_2SO_4.

Decrease of soil acidity

Similarly as described above, the pH was increased by one unit using $Ca(OH)_2$.

Addition of lead

The soils were contaminated with different quantities of lead (Pb), namely 50; 500; 1000 and 5000 mg/kg. The lead was added as $PbCl_2$, this being one of the main lead salts in exhaust fumes from petrol engines (Brunner, 1966).

Addition of pyrazon

The soils were treated with the herbicide pyrazon (5-amino-4-chloro-2-phenylpyridazin-3(2H)-one). The treatment rates corresponded to 1x and 10x the recommended field application, i.e. 2.5 kg a.i./ha and 25 kg a.i./ha. The powdered formulation pyramin (80% pyrazon) was used. For the laboratory experiments, conversion of field application rates to mg herbicide per kg soil was calculated assuming an even distribution of the pesticide in the plough layer (4.5 million kg soil/ha). This is a simplification of the natural

148

Table 1   Physico-chemical properties of the various soil types

Soil name	Texture	Vegetational cover of land use	pH water	pH KCl	Mineral fraction (%)			Org. C (%)	Kj-N (%)	C.E.C. meq/ 100 g dry soil
					Clay <2µm	Silt 2-50µm	Sand >50µm			
1. Gistel	sandy clay	grassland	6.5	6.2	21.4	23.9	54.7	4.01	0.274	14.4
2. Bierbeek	fine sandy	grassland	5.6	5.0	6.1	31.6	62.3	2.24	0.142	6.1
3. Bonheiden	sand	natural vegetation	4.3	3.9	0.6	0.8	98.6	0.11	0.013	2.7

situation where a concentration gradient, determined by the nature of the pesticide, the soil, the spray volume, and so on, may exist.

The treated soils, and a non–treated control soil, were stored in plastic bags and incubated as described above. The moisture content was controlled during the whole experiment, being adjusted when necessary to 75% of the field moisture capacity. After 2, 4, 12 and 24 weeks, samples of 100 g wet weight of soil were taken for biokinetic analysis.

For the treated and control soils the biological oxygen consumption was measured in the Sapromat respirometer (Voith, W. Germany). Of each soil treatment, subsamples of 100 g wet weight were brought in the Sapromat Erlenmeyer flask and amended with the following nutrients: glucose monohydrate 120 mg; yeast extract 30 mg; NH_4Cl 45 mg; $MgSO_4.7H_2O$ 12 mg and KH_2PO_4 10 mg. The Erlenmeyers were incubated in the respirometer at 20°C at a constant oxygen partial pressure of 20%. The amount of oxygen consumed was registered continuously over a period of 5 days (120 hours). In addition, for the different soils at the different stages, the values for u_{max} and K_S were determined in separate experiments. To a series of soil samples, increasing amounts of glucose-medium were added and the resulting rates of oxygen consumption were monitored in the respirometers. A double reciprocal plot of the rate of oxygen consumption versus substrate concentration (Lineweaver–Burk transformation) analysed by linear regression gave values for μ_{max} and K_S. Hence, the biokinetic analysis of the 5 day respiration curves with u_{max}, K_S and S_O as independent inputs. From the curve, mathematical simulation yielded an estimate of X_0, Y_{max} and m. For further details, the reader is referred to Van de Werf and Verstraete (1986a,b).

RESULTS

Influence of pH changes

The soil pH was respectively increased and decreased by one unit. The influence of these pH changes on the biokinetic parameters of the tested soils are summarized in Tables 2, 3 and 4.

Table 2   Influence of pH changes and storage times on the biokinetic parameters of the Gistel soil

TIME (weeks)	CONTROL (A)					pH + 1 (B)					pH - 1 (C)				
	μ_{max}	K_S	y_{max}	m	X_0	μ_{max}	K_S	y_{max}	m	X_0	μ_{max}	K_S	y_{max}	m	X_0
0	0.1617	379	0.545	0.0115	142.6	0.1617	379	0.545	0.0115	142.6	0.1617	379	0.545	0.0115	142.6
2	0.1770	82	0.682	0.0225	75.2	0.2894	136	0.721	0.0255	73.8	0.1938	761	0.525	0.0108	34.7
4	0.1890	200	0.625	0.0155	24.4	0.2151	247	0.652	0.0225	86.2	0.1957	436	0.539	0.0125	27.2
12	0.1991	154	0.592	0.0152	19.8	0.2688	191	0.689	0.0220	53.4	0.2646	689	0.607	0.0184	22.8
24	0.2871	780	0.602	0.0152	12.5	0.2496	885	0.581	0.0196	39.2	0.2158	395	0.615	0.0198	24.8

Biokinetic parameter	μ_{max}	K_S	y_{max}	m	X_0
Treatment*	A C B	A B C	C A B	C A B	C A B
Duncan-test					
P: 0.01	-------	-------	-------	-------	-------
P: 0.05	-------	-------	-------	-------	-------
L.S.D. (with regard to control)					
P: 0.01	0.074	532.721	0.110	0.008	43.374
P: 0.05	0.051	365.615	0.076	0.006	29.769
Coefficient of variation (%)	16.249	61.738	8.601	23.270	26.836

* Arranged in increasing order

151

Table 3   Influence of pH changes and storage times on the biokinetic parameters of the Bierbeek soil

TIME (weeks)	CONTROL (A)					pH + 1 (B)					pH – 1 (C)				
	μ_{max}	K_S	y_{max}	m	x_0	μ_{max}	K_S	y_{max}	m	x_0	μ_{max}	K_S	y_{max}	m	x_0
0	0.2175	1557	0.472	0.0034	34.5	0.2175	1557	0.472	0.0034	34.5	0.2175	1557	0.472	0.0034	34.5
2	0.2210	1819	0.445	0.0064	67.2	0.2118	1738	0.446	0.0081	38.5	0.2236	1677	0.482	0.0109	33.7
4	0.2325	1922	0.452	0.0078	56.8	n.d.	n.d.	n.d.	n.d.	n.d.	0.3317	2434	0.497	0.0117	22.8
12	0.2432	2015	0.427	0.0086	43.2	0.2523	2645	0.421	0.0074	59.6	0.2616	2745	0.452	0.0094	46.4
24	0.3068	3063	0.436	0.0082	36.6	0.2420	3030	0.392	0.0062	52.2	0.3839	3264	0.505	0.0126	46.5

Biokinetic parameter	μ_{max}	K_S	y_{max}	m	x_0
Treatment*					
Duncan-test	B A C	A B C	B A C	B A C	C B A
P: 0.01	————	————	————	————	————
P: 0.05	————	————	————	————	————
L.S.D. (with regard to control)					
P: 0.01	0.080	496.485	0.051	0.003	33.416
P: 0.05	0.054	316.785	0.034	0.002	22.547
Coefficient of variation (%)	14.251	9.573	5.023	19.718	30.010

* Arranged in increasing order

n.d.: not determined

152

Table 4   Influence of pH changes and storage times on the biokinetic parameters of the Bonheiden soil

TIME (weeks)	CONTROL (A)					pH + 1 (B)					pH – 1 (C)				
	μ_{max}	K_s	ymax	m	X_0	μ_{max}	K_s	ymax	m	X_0	μ_{max}	K_s	ymax	m	X_0
0	0.1364	332	0.494	0.0015	2.8	0.1364	322	0.494	0.0015	2.8	0.1364	322	0.494	0.0015	2.8
2	0.1341	155	0.558	0.0095	2.1	0.1004	622	0.482	0.0072	3.7	0.1356	886	0.521	0.0185	2.3
4	0.1728	663	0.562	0.0093	1.7	0.1317	1176	0.518	0.0076	1.8	0.1685	282	0.685	0.0225	0.4
12	0.1200	257	0.552	0.0081	0.7	0.1264	855	0.515	0.0042	1.1	n.d.	n.d.	n.d.	n.d.	n.d
24	0.1065	632	0.572	0.0032	0.5	0.0986	530	0.541	0.0062	0.7	0.2736	549	0.595	0.0215	0.2

Biokinetic parameter	μ_{max}			K_s			ymax			m			X_0	
Treatment*	B	A	C	A	C	B	B	A	C	B	A	C	A	B
Duncan-test														
P: 0.01	————			————			————			————			————	
P: 0.05	————			————			————			————			————	
L.S.D. (with regard to control)														
P: 0.01	0.098			668.561			0.009			0.087			1.044	
P: 0.05	0.066			451.110			0.006			0.059			0.704	
Coefficient of variation (%)	30.880			56.025			45.162			7.224			28.857	

* Arranged in increasing order
n.d.: not determined

In the biomass-rich Gistel soil, the control reveals that storage strongly decreases the active biomass. In addition, a shift towards a community with an increased specific growth rate but decreased substrate affinity is noticeable. When the pH was raised, similar trends can be observed, although the decrease of the active biomass is not so pronounced. In the acidified soil, X_0 first decreases very rapidly, but then stabilizes at a higher level than in the control. For the whole of the time course, the parameters μ_{max}, K_s, Y_{max} and m of the treated soils did not differ (P=0.01) for those of the control soil. The parameter X_0 appeared to be increased significantly (P=0.05) as a result of the addition of lime. Acidification had no significant impact on the active soil biomass.

For the Bierbeek soil, the control soil first increased in active biomass, but upon further storage decreased to about half of the maximum obtained. The trends for μ_{max} and K_s were similarly to those for the Gistel soil. Alteration of the pH appeared to have only an effect on the substrate affinity and the maintenance coefficient. The acid podzol Bonheiden behaved quite similarly to Gistel. Active biomass dropped considerably upon storage. The increase or decrease of the pH affected the X_0 and m parameter only.

Influence of lead pollution

The active soil biomass values, and their biokinetic properties, are represented in Tables 5, 6 and 7. Statistical analysis indicates that none of the biokinetic parameters is significantly altered at lead concentrations below 50 ppm. The exception to this is the podzolic Bonheiden soil for which the m parameter reveals significant differences between A and B. Lead levels above 1000 ppm give rise to a significant decrease in the level of the active microbial biomass in all three soils tested. Indeed, all treatments E are significantly different from A for the X_0 parameter (P=0.05).

Influence of pyrazon

The results for the Gistel, Bierbeek and Bonheiden soils are given in Tables 8, 9 and 10 respectively. For this chemical, treatments A and B were different (P=0.05) for Y_{max} in the Gistel soil and for X_0 in the Bonheiden soil. At the dose of 25 kg a.i./ha, there were differences significant (P=0.05) for Y_{max} and m in the

Gistel soil, for Ymax in the Bierbeek soil, and for μmax in the Bonheiden soil. No significant differences were detected for XO, except a minor stimulation for the podzol at the lower herbicide level.

DISCUSSION

For the soils tested, the active microbial biomass differs greatly with respect to μmax, Ks, m and the total level. Yet, in these three soils, the impact of the various treatments on the active microbial biomass is quite similar.

The most important effect noticed is that of storage. A shift towards a community with a decreased active soil biomass which, furthermore, has a lower substrate affinity, is observed in all three soils. The reason why, in the Bierbeek soil, the active biomass first increased and only then decreased is not clear, but is probably due to the presence of easily degradable substrates. It should be noted that the effect of storage on a variety of other soil microbial parameters such as mineralized nitrogen flush; CO_2-production and ATP content, has been studied by Ross et al., (1980). None of these methods was capable of revealing such a strong shift in the microbial composition as this biokinetic analysis.

Statistical analysis reveals that storage brings about a significant linear decrease of the active soil biomass (XO) and a highly significant linear increase of the substrate affinity coefficient (Ks). The maximum specific growth rate (μmax) is not effected by storage (Table 11).

The alteration of soil pH is normally considered to have a considerable influence on the soil microbial community (Sparling and Cheshire, 1979; Francis et al., 1980 and Lohm et al., 1984). The method used in this work does not detect these alterations in the soils examined. Only liming has a significantly positive effect on the Gistel soil.

The addition of lead is generally considered to affect the microbial community of soils when present at levels of 375 mg lead/kg and onwards (Doelman and Haanstra, 1979). In our experiments, significant effects on the active soil biomass were registered for the Bierbeek soil at 50 ppm, for the Bonheiden soil at 1000 ppm and for the Gistel at 5000 mg lead/kg.

The effect of pyrazon on the soil microbiota is generally found to be insignificant at application levels

Table 5   The effect of various Pb concentrations (added as PbCl$_2$) on the biokinetic parameters of the Gistel soil

TIME (weeks)	CONTROL (A)					50 ppm Pb (B)					500 ppm Pb (C)				
	μ_{max}	K_s	γ_{max}	m	x_0	μ_{max}	K_s	γ_{max}	m	x_0	μ_{max}	K_s	γ_{max}	m	x_0
0	0.1617	379	0.545	0.0115	142.6	0.1617	379	0.545	0.0115	142.6	0.1617	379	0.545	0.0115	142.6
2	0.1770	82	0.682	0.0225	75.2	0.1667	398	0.569	0.0125	28.2	0.1707	576	0.605	0.0145	43.2
4	0.1890	200	0.625	0.0155	24.4	0.1783	558	0.525	0.0118	21.4	0.2258	612	0.559	0.0092	25.2
12	0.1991	154	0.592	0.0152	19.8	0.1842	549	0.541	0.0085	19.9	0.2202	48	0.618	0.0225	9.8
24	0.2871	780	0.602	0.0152	12.5	0.1880	671	0.527	0.0102	18.9	0.2105	191	0.531	0.0108	8.5

TIME (weeks)	1000 ppm Pb (D)					5000 ppm Pb (E)				
	μ_{max}	K_s	γ_{max}	m	x_0	μ_{max}	K_s	γ_{max}	m	x_0
0	0.1617	379	0.545	0.0115	142.6	0.1617	379	0.545	0.0115	142.6
2	0.2741	644	0.645	0.0140	13.2	0.3148	3485	0.427	0.0095	8.0
4	0.4694	1808	0.582	0.0207	7.9	0.2003	293	0.544	0.0086	2.0
12	0.3979	524	0.533	0.0075	1.8	0.2808	769	0.526	0.0123	1.1
24	0.1901	733	0.605	0.0178	45.2	0.2801	705	0.585	0.0163	2.5

156

Biokinetic parameter	μ_{max}	K_s	y_{max}	m	x_0
Treatment* Duncan-test	B C A E D	A C B D E	E B C D A	B E C D A	E D C B A
P: 0.01					
P: 0.05					
L.S.D. (with regard to control)					
P: 0.01	0.123	1274.500	0.086	0.008	33.062
P: 0.05	0.089	925.787	0.063	0.006	24.016
Coefficient of variation (%)	29.681	110.054	8.264	32.617	32.809

* Arranged in increasing order

157

Table 6   The effect of various pb concentrations (added as PbCl2) on the Bierbeek soil

TIME (weeks)	CONTROL (A)					50 ppm Pb (B)					500 ppm Pb (C)				
	μ_{max}	K_s	y_{max}	m	X_0	μ_{max}	K_s	y_{max}	m	X_0	μ_{max}	K_s	y_{max}	m	X_0
0	0.2175	1557	0.472	0.0034	34.5	0.2175	1557	0.472	0.0034	34.5	0.2175	1557	0.472	0.0034	34.5
2	0.2210	1819	0.445	0.0064	67.2	0.2313	1617	0.435	0.0092	26.2	0.2426	1695	0.435	0.0125	25.4
4	0.2325	1922	0.452	0.0078	56.8	0.3020	1331	0.492	0.0097	23.2	0.2816	1684	0.464	0.0085	18.3
12	0.2432	2015	0.427	0.0086	43.2	0.3169	3565	0.379	0.0064	30.6	0.3230	3295	0.417	0.0119	24.5
24	0.3068	3063	0.436	0.0082	36.6	0.3170	3610	0.402	0.0064	28.8	0.3515	3448	0.412	0.0078	20.5

TIME (weeks)	1000 ppm Pb (D)					5000 ppm Pb (E)				
	μ_{max}	K_s	y_{max}	m	X_0	μ_{max}	K_s	y_{max}	m	X_0
0	0.2175	1557	0.472	0.0034	34.5	0.2175	1557	0.472	0.0034	34.5
2	0.5486	773	0.480	0.0167	0.5	0.2585	428	0.556	0.0079	0.6
4	0.4224	1508	0.549	0.0155	9.9	0.1847	52	0.572	0.0058	0.8
12	0.3109	1264	0.512	0.0076	5.4	0.2374	886	0.552	0.0105	0.1
24	0.2918	817	0.556	0.0082	2.6	0.1968	1051	0.500	0.0083	0.4

Biokinetic parameter	μ_{max}	K_s	y_{max}	m	x_0
Treatment* Duncan-test	E A B C D	E D A C B	E C A D E	A B E C D	E D C B A
P: 0.01					
P: 0.05					
L.S.D. (with regard to control)					
P: 0.01	0.126	1185.818	0.063	0.005	23.246
P: 0.05	0.092	861.369	0.046	0.003	16.885
Coefficient of variation (%)	24.771	36.790	7.226	31.050	45.664

* Arranged in increasing order

Table 7   The effect of various Pb concentrations (added as PbCl2) on the biokinetic parameters of the Bonheiden soil

TIME (weeks)	CONTROL (A) μ_{max}	K_s	y_{max}	m	X_0	50 ppm Pb (B) μ_{max}	K_s	y_{max}	m	X_0	500 ppm Pb (C) μ_{max}	K_s	y_{max}	m	X_0
0	0.1364	322	0.494	0.0015	2.8	0.1364	322	0.494	0.0015	2.8	0.1364	322	0.494	0.0015	2.8
2	0.1341	155	0.558	0.0095	2.1	0.0954	1083	0.425	0.0025	1.8	0.1082	1320	0.415	0.0015	1.7
4	0.1728	663	0.562	0.0093	1.7	0.1312	1193	0.492	0.0025	0.8	0.1214	1245	0.475	0.0010	0.7
12	0.1200	257	0.552	0.0081	0.7	0.1419	1182	0.478	0.0035	0.6	0.1030	1160	0.492	0.0000	1.4
24	0.1065	632	0.572	0.0032	0.5	0.1566	1475	0.492	0.0025	0.4	0.1006	1254	0.325	0.0000	0.4

TIME (weeks)	1000 ppm Pb (D) μ_{max}	K_s	y_{max}	m	X_0	5000 ppm Pb (E) μ_{max}	K_s	y_{max}	m	X_0
0	0.1364	322	0.494	0.0015	2.8	0.1364	322	0.494	0.0015	2.8
2	0.0811	3747	0.0	0.0000	0.1	0.0906	1514	0.435	0.0000	0.3
4	0.1912	4055	0.062	0.0000	0.1	0.1339	1536	0.659	0.0000	0.4
12	0.1081	1252	0.585	0.0000	0.2	0.0480	334	0.136	0.0000	0.2
24	0.1340	4395	0.003	0.0000	0.0	0.1270	1338	0.315	0.0000	0.5

Biokinetic parameter	μ_{max}	K_s	y_{max}	m	x_0
Treatment* Duncan-test	E C D B A	A E B C B	D E C B A	E D C B A	D E B C A
P: 0.01					
P: 0.05					
L.S.D. (with regard to control)					
P: 0.01	0.044	1330.542	0.302	0.003	0.822
P: 0.05	0.032	966.496	0.219	0.002	0.597
Coefficient of variation (%)	19.466	57.355	38.837	89.544	38.291

* Arranged in increasing order

Table 8   Influence of pyrazon concentrations and storage time on the biokinetic parameters of the Gistel soil

TIME (weeks)	CONTROL (A)					PYRAZON: 2.5 kg a.i./ha (B)					PYRAZON: 25 kg a.i./ha (C)				
	μ_{max}	K_S	y_{max}	m	X_0	μ_{max}	K_S	y_{max}	m	X_0	μ_{max}	K_S	y_{max}	m	X_0
0	0.1617	379	0.545	0.0115	142.6	0.1617	379	0.545	0.0115	142.6	0.1617	379	0.545	0.0115	142.6
2	0.1770	82	0.682	0.0225	75.2	0.1606	39	0.591	0.0158	22.4	0.1533	263	0.585	0.0137	40.2
4	0.1890	200	0.625	0.0155	24.4	0.2013	638	0.592	0.0142	20.4	0.1636	658	0.558	0.0097	39.3
12	0.191	154	0.592	0.0152	19.8	0.2093	623	0.578	0.0157	34.2	0.2630	1344	0.559	0.0078	30.1
24	0.2871	780	0.602	0.0152	12.5	0.1889	377	0.592	0.0118	31.8	0.2346	749	0.582	0.0098	27.5

Biokinetic parameter	μ_{max}			K_S			y_{max}			m			X_0		
Treatment*	B	C	A	A	B	C	C	B	A	C	B	A	B	A	C
Duncan-test	———		——	———		——	———		——	———		——	———		——
L.S.D. (with regard to control)															
P: 0.01	0.064			597.360			0.047			0.005			40.189		
P: 0.05	0.044			409.978			0.033			0.003			27.582		
Coefficient of variation (%)	15.627			59.882			3.816			16.381			28.447		

* Arranged in increasing order

162

Table 9   Influence of pyrazon concentrations and storage time on the biokinetic parameters of the Bierbeek soil

TIME (weeks)	CONTROL (A)					PYRAZON: 2.5 kg a.i./ha (B)					PYRAZON: 25 kg a.i./ha (C)				
	μ_{max}	K_S	ymax	m	X_0	μ_{max}	K_S	ymax	m	X_0	μ_{max}	K_S	ymax	m	X_0
0	0.2175	1557	0.472	0.0034	34.5	0.2175	1557	0.472	0.0034	34.5	0.2175	1557	0.472	0.0034	34.5
2	0.2210	1819	0.445	0.0064	67.2	0.2101	1724	0.414	0.0069	27.8	0.2458	2694	0.409	0.0079	37.4
4	0.2325	1922	0.452	0.0078	56.8	0.2577	2475	0.398	0.0052	30.2	n.d.	n.d.	n.d.	n.d.	n.d
12	0.2432	2015	0.427	0.0086	43.2	0.2208	2225	0.418	0.0076	38.0	0.2214	1946	0.402	0.0086	22.2
24	0.3068	3063	0.436	0.0082	36.6	0.2443	2763	0.428	0.0098	32.5	0.2347	2568	0.409	0.0080	31.1

Biokinetic parameter	μ_{max}	K_S	ymax	m	X_0
Treatment*	B C A	A B C	C B A	B A C	C B A
Duncan-test					
P: 0.01	------	------	------	------	------
P: 0.05	------	------	------	------	------
L.S.D. (with regard to control)					
P: 0.01	0.052	771.701	0.029	0.002	25.946
P: 0.05	0.035	520.704	0.020	0.001	17.507
Coefficient of variation (%)	9.905	16.247	3.040	14.554	26.972

* Arranged in increasing order

n.d. = not determined

163

Table 10 Influence of pyrazon concentrations on the biokinetic parameters of the Bonheiden soil

TIME (weeks)	CONTROL (A)					PYRAZON: 2.5 kg a.i./ha (B)					PYRAZON: 25 kg a.i./ha (C)				
	μ_{max}	K_S	y_{max}	m	x_0	μ_{max}	K_S	y_{max}	m	x_0	μ_{max}	K_S	y_{max}	m	x_0
0	0.1364	322	0.494	0.0015	2.8	0.1364	322	0.494	0.0015	2.8	0.1364	322	0.494	0.0015	2.8
2	0.1341	155	0.558	0.0095	2.1	0.1579	278	0.555	0.0068	4.3	0.1726	188	0.567	0.0118	2.1
4	0.1728	663	0.562	0.0093	1.7	0.1682	476	0.531	0.0054	3.8	0.2335	846	0.545	0.0065	1.7
12	0.1200	257	0.552	0.0081	0.7	0.1784	424	0.565	0.0086	3.6	0.1784	299	0.553	0.0091	1.8
24	0.1065	632	0.572	0.0032	0.5	0.1645	1239	0.564	0.0035	0.4	0.1521	974	0.587	0.0068	1.1

Biokinetic parameter	μ_{max}	K_S	y_{max}	m	x_0
Treatment*	A B C	A C B	B A C	B A C	A C B
Duncan-test					
P: 0.01	⌐ ¬	⌐ ¬	⌐ ¬	⌐ ¬	⌐ ¬
P: 0.05	⌐ ¬	⌐ ¬	⌐ ¬	⌐ ¬	⌐ ¬
L.S.D. (with regard to control)					
P: 0.01	0.043	351.027	0.021	0.003	1.788
P: 0.05	0.029	240.915	0.014	0.002	1.227
Coefficient of variation (%)	12.910	33.509	1.814	24.958	38.193

* Arranged in increasing order

Table 11   Variation of the biokinetic parameters μ_{max}, K_s and the active soil biomass X_0 as a function of the storage time (parameters are expressed in %, initial values X_0 = 100 %)

Time (t) (weeks)	μ_{max}			K_s			X_0		
	Gistel	Bierbeek	Bonheiden	Gistel	Bierbeek	Bonheiden	Gistel	Bierbeek	Bonheiden
0	100	100	100	100	100	100	100	100	100
2	109	101	98	21	48	48	195	195	75
4	116	106	126	52	205	205	164	164	61
12	123	111	87	40	79	79	125	125	25
24	177	141	78	205	196	196	106	106	18
Reg. analysis		sum of squares	F calc		sum of squares	F calc		sum of squares	F calc
Linear component of regression		1474	2.86		20208	11.91**		7633	7.52*
Quadratic compt of regression		49	0.09		5707	3.36		797	0.78
Cubic component of regression		377	0.73		1725	1.01		65	0.06
Rest component Regression		120	-		6337			565	
					$K_s = 79 + 4.1\,t$			$X_0 = 99 - 2.5\,t$	
Coefficient of variation (%)		20			35			41	

* Significant difference at the 5% level
** Significant difference at the 1% level

165

corresponding to both normal field rates, and rates ten times higher than that (Engvild and Jensen 1969; Fröhner et al., 1970). Our experiments indicate that, indeed, this chemical at these levels does not cause consistent alteration of the biokinetic characterics of the soil microbiota.

Of the three soils examined, the podzol Bonheiden intuitively qualifies as a rather weak ecosystem. The Bierbeek soil represents a normal ecosystem while the Gistel soil, by its high level of organic carbon and C.E.C., represents a well buffered ecosystem. The overview of the impact of the different treatments on the active soil microbial biomass, as schematized in Table 12, indicates that indeed the first two soils are somewhat more sensitive than the rich Gistel soil.

Table 12   Overview of the impact of the different treatments on the active soil microbial biomass X_0

Treatment	Type of soil	Podzol 0.11% C (Bonheiden)	Sandy loam 2.2% C (Bierbeek)	Sandy clay loam 4.0% C (Gistel)
pH	+ 1 unit	NS	NS	O
	- 1 unit	NS	NS	NS
Pb	50 ppm	NS	o	NS
	500 ppm	NS	oo	NS
	1000 ppm	oo	oo	NS
	5000 ppm	o	oo	O
Py-ra-zon	0.6 mg a.i./kg	O	NS	NS
	6.0 mg a.i./kg	NS	NS	NS

o   Significant decrease at the 5% level
oo   Significant decrease at the 1% level
O   Significant increase at the 5% level
NS   No significant effect with regard to the control

The sharp decline of the active soil microbial biomass upon storage time substantiates the concern expressed by Anderson <u>et</u> <u>al</u>. (1981) (see also Chapter 3) with regard to the use of stored soils to study side effects of pesticides.

CONCLUSIONS

The overall behaviour of this active soil biomass component to various treatments, and particularly its susceptibility to soil storage indicates that it is a parameter with potential value for microbial ecologists.

APPENDIX

Definitions and units

μ_{max} maximum growth rate $= \dfrac{dX}{dt} \cdot \dfrac{1}{X}$ $(hr-1)$

K_s substrate affinity constant (mg COD/kg soil wet weight)

S_0 substrate concentration at the onset of the respiration test, expressed as chemical oxygen demand COD (mg COD/kg soil wet weight)

X_0 viable biomass, capable to directly start to metabolize the substrate added (mg biomass DW/kg soil wet weight)

Y_{max} maximum yield factor (mg biomass DW formed/mg COD utilised)

m cell maintenance coefficient (mg COD utilised/mg biomass DW hour).

REFERENCES

Anderson, J.P.E. and Domsch, K.H., 1978, Soil Biology and Biochemistry 10, 215.

Anderson, J.P.E. and Domsch, K.H., 1980, Soil Science 130, 211.

Anderson, J.P.E., Armstrong, R.A. and Smith, S.N., 1981, Soil Biology and Biochemistry 13, 149.

Brunner M., 1966, Zeitschrift für Präventivmedizin 11, 77.

Doelman, P. and Haanstra, L., 1979, Soil Biology and Biochemistry 11, 475.

Engvild, K.C. and Jensen, H.L., 1969, Soil Biology and Biochemistry 1, 295.

Francis, A.J., Olsen, D. and Bernatsky, R., 1980, in Ecological Impact of Acid Precipitation, edited by D. Drablos and A. Tollan, (SNSF Project, Norway), p 166.

Fröhner, C., Oltmanns, O. and Lingens, F., 1970, Archiv. für Mikrobiologie 74, 82.

Maanstra, L. and Doelman, P., 1984, Soil Biology and Biochemistry 16, 595.

Jenkinson, D.S. and Powlson, D.S., 1976, Soil Biology and Biochemistry 8, 209.

Ladd, J.N., Oades, J.M. and Amato, M., 1981, Soil Biology and Biochemistry 13, 119.

Lohm, U., Larsson, K. and Nômmik, H., 1984, Soil Biology and Biochemistry 16, 343.

Marumoto, T., Anderson, J.P.E. and Domsch K.H., 1982, Soil Biology and Biochemistry 14, 469.

Paul, E.A. and Johnson, R.L., 1977, Applied and Environmental Microbiology 34, 263.

Ross, D.J., Tate, K.R., Cairns, A. and Meyrick, K.F., 1980, Soil Biology and Biochemistry 12, 369.

Sparling, G.P. and Cheshire, M.V., 1979, Soil Biology and

Biochemistry 11, 317.

Sparling, G.P., 1981, Soil Biology and Biochemistry 13, 93.

Van de Werf, H. and Verstraete W., 1987a, Soil Biology and Biochemistry in press.

Van de Werf, H. and Verstraete W., 1987b, Soil Biology and Biochemistry in press.

Verstraete, W., Van de Werf, H., Kucnerowicz, F., Ilaiwi, M., Verstraeten, L.M.J. and Vlassak, K., 1983, Soil Biology and Biochemistry 15, 391.

Voroney, R.P. and Paul, E.A., 1984, Soil Biology and Biochemistry 16, 9.

CHAPTER 12

RADIORESPIROMETRY MEASUREMENTS
OF MICROBIAL RESPONSE

G. Soulas & J. C. Fournier

INTRODUCTION

One of the main concerns in agricultural use of pesticides is to assess their ecotoxicological impact on organisms, especially microorganisms living in soil. At the present state of investigation a strategy has been developed that proceeds in two steps.

1 : define a set of convenient microbial populations and/or functions that can be used to estimate such effects.
2 : define a framework for interpreting quantitative results given by the test in terms of tolerance by comparing them to those observed after natural stresses (Greaves et al., 1980; Greaves and Malkomes, 1980).

Perhaps the main problem in applying such a strategy resides in the difficulty of finding an appropriate experimental approach. This must be accurate and sensitive enough and should have characteristics of specificity that make it useful for following the responses of broad or specific microbial groups.
In this paper we describe a method for studying side-effects of pesticides on the soil microflora which is based on in situ labelling of soil microorganisms. Because it can be easily performed such a method might be proposed as a convenient test.

DESCRIPTION OF THE METHOD

The method is based on in situ labelling of groups of microorganisms that may be representative, either of the total soil microflora or, in contrast to this, of more restricted microbial populations such as specific, physiological groups. We proceed in two successive steps.

Firstly, the soil microflora, or a part of it, is labelled by adding appropriate ^{14}C-substrates to soil samples. Two types of substrates have been chosen for this first experiment: ^{14}C-glucose that is largely used as a carbon-source by the main fraction of the soil microflora and ^{14}C-2,4-D that is assimilated by a specific and small microbial population of the soil.

After the ^{14}C-substrate is assumed to have been completely assimilated, as indicated by a slowing down in the rate of $^{14}CO_2$ evolution, non-labelled biologically active compounds are added to the soil samples. Subsequent modifications in the $^{14}CO_2$ evolution kinetics are then followed. Chloroform fumigated samples give a measurement of the total radioactivity that has been incorporated into the biomass.

EXPERIMENTAL PROCEDURE

The soil characteristics were: loam 54%, clay 32%, silt 13%, organic carbon (C) 1.2%, pH 7.4% (water), moisture handling capacity (WHC) 26%. The soil moisture was maintained at 75% of WHC during the first step of the experiment (labelling) and was 80% of WHC during the second step (toxicity determination). Soil samples of 50 g (dry weight) were incubated in two litre glass jars at 20°C.

First step: labelling of the soil microorganisms

Two types of experiments have been conducted with ^{14}C-glucose. In the first, amounts of glucose as high as 1000 mg (carbon)/kg (soil) were added to the soil samples. (NH4)2SO4 was also added as a nitrogen source to get a C/N ratio of 10. In a second experiment only minute amounts of glucose representing 1 mg (C)/kg (soil) were added to the soil samples.

Each procedure has its own advantages. As indicated by recent experiments (Fournier and Chaussod, unpublished results), using small amounts of glucose allows the major part of the radioactivity to be quickly and predominantly incorporated into the soil microflora. As a main consequence the second step of the experiment, the "toxicity" assay, may be performed only one week after glucose application. In contrast use of, higher concentrations of glucose requires a longer time (at

least 2 months) to reach the necessary stabilization in
$14CO_2$ evolution. On the other hand a greater
percentage of radioactivity was evolved as $14CO_2$. So
a lesser percentage of radioactivity may be expected to
be incorporated into the soil microflora. Such a
procedure gives probably a more uniform distribution of
the radioactivity between microbial species and between
microbial constituents. However a question remains to be
answered: what part of the radioactivity which is located
on non-living soil organic carbon is still available to
soil microbes?

The second substrate we have worked with is
$14C-2,4-D$ which is uniformly ring-labelled. Labelling
of the soil microflora was made at 3 mg (2,4-D)/kg
(soil). As with high concentrations of glucose, at least
two months are necessary before stabilization of
$14CO_2$ evolution is reached.

Second step: "toxicity" experiment

Six days after addition of the lower concentration
of glucose and 60 days in all other cases (indicated by
an arrow on Figure 1), different bioactive compounds are
added to the radioactive soil samples. Bioactive

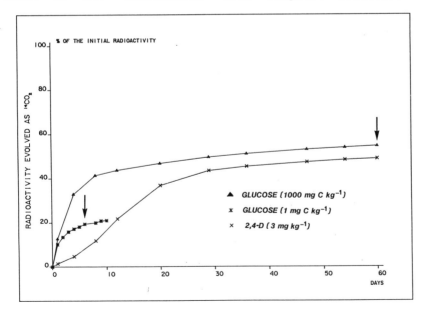

% OF THE INITIAL RADIOACTIVITY

RADIOACTIVITY EVOLVED AS $^{14}CO_2$

▲ GLUCOSE (1000 mg C kg^{-1})
× GLUCOSE (1 mg C kg^{-1})
× 2,4-D (3 mg kg^{-1})

DAYS

Figure 1   Kinetics of $14CO_2$ evolution from the $14C$-substrates
used for labelling soil biomass

compounds are either common microbial C-substrates: glucose at 1000 or 100 mg (C)/kg soil, benzoate at 100 mg (C)/kg or pesticides: 2,4-D at 3 or 4 or 30 mg/kg, DNOC at 3 or 4 mg/kg, CIPC at 4 mg/kg, dichlobenil at 4 and 16 mg/kg, 2,4,5-T at 3 mg/kg and benomyl (as benlate) at 3 mg (a.i.)/kg. After addition of these compounds the kinetics of $^{14}CO_2$ release are followed. Some radioactive soil samples were left without any addition of bioactive compounds as controls.

RESULTS

Comparison between the mineralization kinetics of the different labelled substrates

Figure 1 gives the respective amounts of radioactivity that were evolved as $^{14}CO_2$: these depend on the radioactive substrate and the rate of application. After 10 days only 21% of the radioactivity has been released as $^{14}CO_2$ after the low level of ^{14}C-glucose application.

After a 60 day incubation period with the higher level of ^{14}C-glucose or with ^{14}C-2,4-D the percentages of radioactivity recovered as $^{14}CO_2$ amount respectively to 54.6% and 49.0% of the initial radioactivity.

Determination of the radioactivity incorporated in the soil biomass

Table 1   Distribution of the radioactivity at the end of the preincubation period

	TREATMENT		
	GLUCOSE 1000 mg C/Kg	GLUCOSE 1 mg C/kg	2,4-D 3 mg/kg
^{14}C EVOLVED AS $^{14}CO_2$	54.6(a)	17.0(b)	49.0(a)
^{14}C IN THE BIOMASS	25.2(c)	72.7(d)	8.6(c)
NON LIVING CARBON	20.2	10.3	42.4

(Figures are expressed as percentages of the initially added radioactivity. (a) after 60 days of incubation; (b) after 6 days of incubation; (c) measured at 20°C; (d) measured at 28°C).

The 14C-activity of the biomass was estimated from the flushes of $14CO_2$ from chloroform fumigated soil samples. The results reported in Table 1 have been estimated by the method of Jenkinson and Powlson (1976) modified by Chaussod and Nicolardot (1982). The part of the $14CO_2$ flush that comes from living carbon is calculated in the following way: total amount of $14CO_2$ liberated during the first seven days (at 28°C) or 14 days (at 20°C) is subtracted by the amount of $14CO_2$ liberated between 7 and 14 days (at 28°C) or 14 and 28 days (at 20°C). We used a K_C value of 0.41 (Anderson and Domsch, 1978).

From Table 1, it appears that radioactivity incorporated into the biomass appears to be more important at a low level of glucose application. Only a minor part of the initial radioactivity (8.6%) may be recovered in the biomass when 2,4-D is used as specific marker.

Kinetics of $14CO_2$ evolution after addition of different bioactive compounds

The radioactivity liberated from the controls is not dependent on the nature and amount of the marker substrate and, as a consequence, not on the radioactivity in the soil biomass (Table 2).

Table 2   Radioactivity evolved from the controls during the toxicity experiment

(Figures are expressed as percentages of the initially added radioactivity).

	TREATMENT		
	GLUCOSE 1000 mg C Kg^{-1}	GLUCOSE 1 mg C kg^{-1}	2,4-D 3 mg kg^{-1}
14C EVOLVED AS $14CO_2$ IN THE CONTROLS	7.2 (after 63 days)	9.5 (after 35 days)	7.5 (after 63 days)

Effects of different bioactive compounds on $14C$-glucose pretreated soil samples

DNOC (4 mg/kg soil) and dichlobenil (16 and 4 mg/kg) are the most effective chemicals in liberating excess of

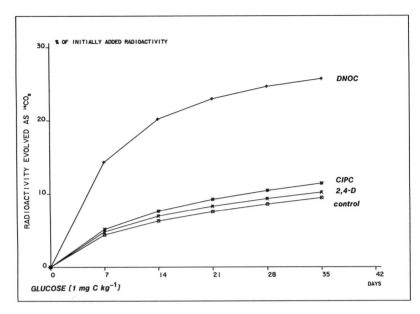

Figure 2a Effect of different bioactive compounds on kinetics of $^{14}CO_2$ evolution from soil samples previously treated with 1 mg ^{14}C–glucose/kg soil

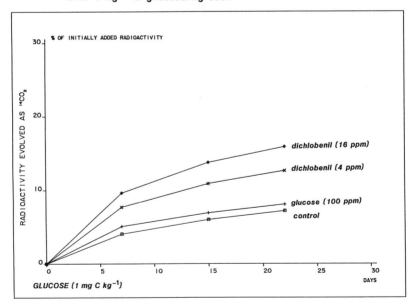

Figure 2b Effect of different bioactive compounds on kinetics of $^{14}CO_2$ evolution from soil samples previously treated with 1 mg ^{14}C–glucose/kg soil

radioactivity from soil samples previously labelled with
^{14}C–glucose at 1 mg (C)/kg (Figures 2a and 2b). The
percentage of radioactivity evolved as $^{14}CO_2$ after
addition of these pesticides reaches 25% with DNOC after
35 days of incubation. It reaches 16% and 12.5% with the
two doses of dichlobenil after 22 days of incubation.
All other compounds, glucose (100 ppm carbon), CIPC
(4 mg/kg soil) and 2,4–D (4 mg/kg) have only limited or
no effects.

The same holds when glucose is used as a marker at
the level of 1000 ppm carbon (Figures 3 and 4). During

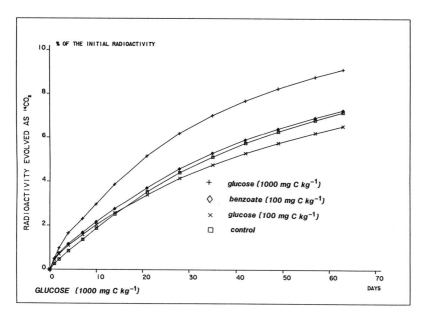

Figure 3   Effect of different common substrates on kinetics
of $^{14}C)_2$ evolution from soil samples previously treated
with 1000 mg^{14}C-glucose/kg soil

the 63 day incubation period, DNOC (3 mg/kg) liberates
8.2% of the initial radioactivity. This corresponds to
18% of the radioactivity present in the soil at the
beginning of the "toxicity" experiment. A second
addition of non–labelled glucose at 1000 ppm carbon also
gives a rise in $^{14}CO_2$ evolution, whereas all other
chemicals have no significant effects.

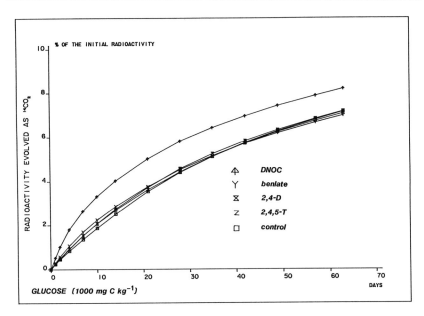

Figure 4   Effect of different pesticides on kinetics of
$^{14}CO_2$ evolution from soil samples previously treated with
1000 mg ^{14}C-glucose/kg soil

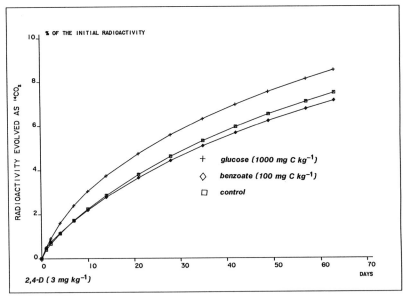

Figure 5   Effect of different common substrates on kinetics of
$^{14}CO_2$ evolution from samples previously treated with 3 mg
2,4-D/kg soil

178

Effects of different bioactive compounds on ^{14}C-2,4-D pretreated soil samples (Figures 5 and 6)

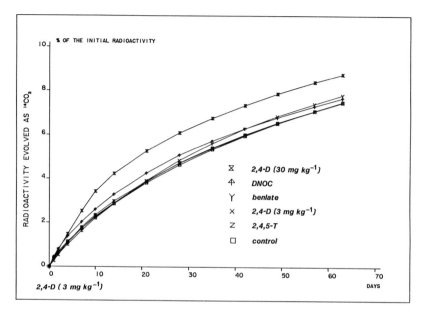

Figure 6   Effect of different chemical on kinetics of $^{14}CO_2$ evolution from soil samples previously treated with 3 mg 2,4-D/kg soil

The most interesting observations are that among the common carbon substrates glucose (1000 ppm) always increases evolution of the remaining soil radioactivity, and that, among the pesticides 2,4-D at 30 mg/kg has an effect exceeding that of DNOC at 3 mg/kg.

DISCUSSION

The method we have developed based on radiorespirometric measurements of soil samples previously amended with ^{14}C-substrates offers two advantages as discussed in the introduction. Sensitivity and adjustable specificity are its main characteristics.

However, the results presented raise essentially two questions. Particularly when the radioactivity located in non-living carbon is important, in what part of the soil carbon does the excess in $^{14}CO_2$ evolution originate after bioactive compounds have been added? Experiments based on the use of very small amounts of ^{14}C-glucose offer a possible explanation. In such

179

cases, radioactivity bound to extracellular metabolites is very limited and so there is some reason to assume that a great part of the radioactivity evolved as $^{14}CO_2$ originates from the living carbon. If such is the case what kind of biological mechanisms are involved to explain such an increase in $^{14}CO_2$ evolution? Figure 7 gives a schematic representation of what could happen when the soil microflora is confronted with biologically active compounds.

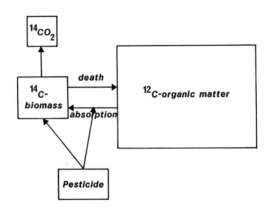

Figure 7   Interaction between soil microflora and biologically active compounds

Common carbon substrates such as glucose and benzoate, are taken up and assimilated by the ^{14}C-labelled biomass for maintenance and, at sufficiently high concentrations, for growth. During the turnover of the cell constituents newly synthesised ^{12}C-compounds replace ^{14}C-compounds that are either degraded within the cell or excreted into the surroundings and re-utilized by other living cells.

On the other hand, some toxic agents might also act as decoupling agents. This is a well-known property of some phenolic compounds. In that case, a greater part of the energy provided by catabolism of soil organic substrates is mainly dissipated as heat. It follows that renewal of cell constituents such as proteins and RNA from exogenous carbon substrates is affected, leading the cells to reutilize part of their intracellular ^{14}C-components.

The third possibility we have to envisage is that some toxic compounds might act as lethal agents, possibly either as a consequence of a prolonged decoupling effect that alters the viability of the cell. Dead microbes are then used as carbon substrates for other living biomass.

REFERENCES

Anderson, J.P.E. and Domsch, K.H., 1978, Soil Biology and Biochemistry, 10, 207.

Chaussod, R. and Nicolardot, B., 1982, Review Ecologie Biologie du Sol, 19, 501.

Greaves, M.P. and Malkomes, H.P., 1980, in Interaction Between Herbicides and the Soil, edited by R.J. Hance (London, Academic Press) p 223.

Greaves, M.P, Poole, N.J., Domsch, K.H., Jagnow, G. and Verstraete, W., 1980. Recommended Tests for Assessing the Side-Effects of Pesticides on the Soil Microflora. Technical Report, Agricultural Research Council, Weed Research Organisation, No. 59.

Jenkinson, D.S. and Powlson, D.S., 1976, Soil Biology and Biochemistry, 8, 209.

CHAPTER 13

SIDE-EFFECTS TESTING:
AN ALTERNATIVE APPROACH

M.P. Greaves

INTRODUCTION

The potential threat to the soil environment posed by the use of pesticides has been recognised for several decades. The imposition by registration authorities of requirements for chemical manufacturers to present data designed to aid prediction of this potential is more recent. Nonetheless, it is now a long-standing requirement of several authorities and a number have considered, the problems and in some cases published guidelines outlining acceptable test protocols.

Despite this considerable international concern, it is striking that there has been little, if any, change in the approach or methods recommended for use. This lack of change is all the more notable in view of the vast amount of soil microbiological literature published in the last twenty-five years.

Perhaps the only change of any significance in the published protocol has been the dropping of tests on pure cultures and soil enzymes by the U.S. Environmental Protection Agency. The EPA's reasons for dropping pure culture and soil enzyme studies are reflected in those given for their omission from published recommended tests by Greaves et al. (1980) and I shall return to these later. This change was followed by the EPA 'reserving' the suggested protocol for testing effects on soil microflora on the grounds that the data obtained in the tests could not be extrapolated to field conditions and, so, was not of any value in predicting likely side-effects. While this author has considerable sympathy with this view, it has to be accepted that some form of testing may be necessary as an overt sign that something is being attempted to try to ensure that the environment is protected.

The acceptance that tests of some kind are necessary involves also the acceptance of the disadvantages that such tests carry. Foremost amongst these is the risk that, by including tests in recommended protocols, it may be accepted by some as a statement that all is well. This conclusion is drawn despite the publication of statements that the protocols only reflect the 'state-of-the-art' and intensive research is required before realistic prediction of side-effects on the field can be made.

It is an absolute requirement of any test method protocol that it should be flexible, "in order that deletion of obsolete methods and introduction of proven improved methods is easily possible" (Greaves et al., 1980). World agriculture will remain dependent on the continuing supply of new and more effective pesticides, despite the problems of overproduction which are current in certain areas. There will, therefore, be a continuing need for evaluation of these chemicals to ensure, as far as possible, that they have no harmful effects on the environment. In accepting this philosophy, however, it is essential to realise that testing costs money. If the costs of implementing environmental safety tests become too great, companies will restrict their production more and more to the relatively few chemicals which have major markets. Almost by definition, these chemicals will be of increasing activity and have broad spectra of activity. Both these factors could confer increased threat to the environment, thus needing more extensive testing and so costing more to develop. At the same time, there is a marked tendency for testing protocols to involve greater use of sophisticated analytical equipment. While this may well give the advantages of greater precision of data and greater throughput of samples via automation, it also demands larger capital expenditure and, more especially, the services of highly paid staff. All these factors significantly increase the costs of testing compounds. In many instances, smaller companies do not have, and cannot afford to establish such facilities and are reliant on the expensive services of contract laboratories. This perpetual vicious cost spiral is already depriving the farmer and grower of much needed tools to manage minority crops and minority pests, weeds and diseases.

Is it possible, therefore, to take a fresh look at the recommended protocols for testing side-effects of pesticides on the soil microflora, with a view to simplifying the system, reducing its cost and yet not losing any value it might have? In this chapter, I shall attempt to do this with the use of some examples from work done at the Weed Research Organization, Oxford, prior to its closure in 1986. In doing this, I wish to express my gratitude to the many members of staff, in particular John Marsh and George Wingfield, who worked with me from 1974 to 1986 and without whom this chapter would not have been possible.

SOME PHILOSOPHIES OF SIDE-EFFECT TESTING

As stated earlier, the basic requirement of any test protocol is that it should be a realistic aid to prediction of likely events in the environment. At the same time, it is necessary that the methods recommended should be sensitive, reproducible and accurate. One could add that they should also be suitable for use in a variety of laboratories with, often, widely different capabilities with regard to equipment and personnel. That is to say, they should be flexible, without losing the essential elements of reproducibility, sensitivity and accuracy. Amongst the reasons given for the reliance of present test protocols on assessment of effects on metabolic processes based on complex analytical techniques, sensitivity and accuracy rate very high. It could be argued, however, that this emphasis is, in some cases at least, too heavy.

The implication is that it is necessary to be able to detect the smallest response to a treatment with a high degree of statistical significance. Is this need so absolute? Professor K.H. Domsch has developed an excellent scheme for interpreting side-effect data (in Greaves et al., 1980) which is universally accepted as a major advance in this area of research. The essence of his scheme is that laboratory-derived data are compared to known data on changes in the soil microflora's activity which result from acceptable or unavoidable natural perturbations such as frost, drought and so on. It is well known that the variability of parameters measured in the field is high (see Cook and Greaves, Chapter 2). Thus, in order to distinguish pesticide-induced effects from this background of high variability they must be large in magnitude and/or duration. Indeed, Professor Domsch's scheme goes on to categorise effects, indicating those which are not tolerable. In this, he states that changes in magnitude (depressions) of fifty percent or more are frequent in nature and so depressions due to man-made chemical stress must not be over- estimated, as long as they are comparable to those found in natural situations. Similarly, duration of effect of 31 to 60 days can, in this scheme, be accepted as tolerable.

It seems, then, that the sensitivity of the method used must only be sufficient to allow detection of changes in control values of a magnitude of two (I assume that some increases can be classed as harmful just as much as decreases). Obviously accuracy is still necessary, as is reproducibility so that the experimental 'reality' of the experimentally observed effect can be judged. That is to say all the numbers derived from the experiments should have some known tolerance attached to them. However, the search for precision must be tempered by the realization that our highly sophisticated analytical capabilities far outreach our abilities to interpret the data in ecological terms.

ALTERNATIVE APPROACHES

If the above arguments are accepted, some considerable opportunities to simplify the test protocols emerge. Perhaps the first phase of simplification is to adopt a clearly tiered system of tests. This is widely practised in many areas of testing pesticides, and, indeed, it has been included in the revised recommendations for microbial testing (see Appendix B). In this instance, however, the tier system starts at an already complex level of test and is based on increasing the range of situations to which the test applies if unacceptable risk is indicated in the first tier. It does nothing to reduce the number of chemicals which must be examined. It is well known that the vast majority of pesticides tested give no effects, or no critical effects, on the parameters measured. Therefore, if these compounds could be detected quickly by a simple, rapid series of tests, that are capable of detecting the large- scale effects discussed earlier, so that only the chemicals likely to give critical effects go forward for testing by more sophisticated methods, the economic savings could be significant. The following examples give one approach that could achieve this.

1)   Pure culture toxicology tests

The use of pure cultures has been proscribed in test

protocols for a variety of reasons. Isolated organisms may be metabolically atypical of their form in soil and may change further in storage. They may be stimulated to artificially high metabolic rates in laboratory media. They are removed from normal ecological associations. Interpretation of results is difficult and extrapolation to the field impossible. These reasons are all valid but similar reasons can show the invalidity of the standard soil incubation techniques which are accepted. Removal of the soil sample from the field and fortification with additives such as powdered plant residue can induce atypical, artificially high metabolic rates, change species balance and so alter normal ecological associations and produce data which is difficult to interpret and impossible to extrapolate to the field. In this latter context, it must be borne in mind that laboratory soil incubations are done using one fixed set of conditions. Variations of soil moisture or temperature, as occur in the field, can markedly change both the rate of reactions, and sometimes their direction, so defeating extrapolation.

A major drawback of pure culture studies proposed in the past as a regulatory test is that they depended on the use of a very few, often only ten or twelve, selected species. Thus, they could in no way be said to approach being representative of the soil population. Further, they usually include only bacteria and fungi, ignoring the major contributions to soil metabolism from actinomycetes, yeasts, algae and protozoa.

The development of miniaturized, and if necessary automated, systems of handling very large numbers of cultures has allowed this disadvantage to be overcome. The systems characteristics also allow large numbers of chemicals to be handled simply. Cooper et al. (1978) have described such a system using plastic microtitration plates as culture vessels for organisms. Each plate contains 96 wells holding up to approximately 0.2ml of culture liquid, and equipment is available to permit simultaneous inoculation of all the wells. Using this system in my laboratory, we have found it easily possible to examine the effects of up to 15 chemicals, each at three concentrations, on some four hundred cultures, including all the major microbial groups.

A particular advantage of the method, apart from the technical ease of handling, is that it appears to be sensitive. This is not surprising as it is accepted that, frequently, micro-organisms are more responsive to chemicals when growing actively than when growing at the rates normally found in soil. To date, nearly seventy pesticides, mainly herbicides, have been examined using this method. The minimum inhibitory concentrations for different micro-organisms have been assessed, as have the proportions of the populations which show inhibited or no growth at set concentrations of chemical.

In the large majority of cases, the toxic effects determined at concentrations ten times higher than that expected in soil were so slight as to cause no concern. This evaluation was confirmed in all cases by comparison with data from conventional side-effect tests. In the few cases where effects on one or more groups of organisms were found, e.g. with glyphosate or paraquat, no significant effects were found in subsequent conventional tests. Thus, this approach certainly seems to satisfy the requirement for sensitivity in that it identifies potential hazard from more chemicals than have been so identified in conventional tests. More important, in one simple, cheap test, the vast majority of chemicals examined were identified as safe and so needed no further expensive testing.

2. Soil tests

Although I feel that the pure culture test, above, is on its own an adequate Tier 1 test to identify compounds needing no further examination, it may be advantageous to have parallel studies of effects _in situ_ in soil. These may be concerned with major metabolic functions. For many years, soil microbiologists have successfully used buried substrate techniques to examine many facets of soil metabolism. Such techniques lend themselves well to assessing pesticide effects, provided the argument that only large effects are significant is accepted.

Organic matter breakdown is still studied by the litter bag technique. Indeed, that technique still forms part of the revised recommendations reproduced in Appendix B to this book. A major problem of this technique, when used for registration purposes, stems from the choice of substrate. It is apparently simple to suggest the use of a plant material relevant to the intended use of the pesticide, e.g. straw for cereal pesticides and apple leaves for those used in orchards. However, it is not so simple as that. There are considerable structural differences between straws from different cereal varieties, let alone species, and these may markedly affect degradation and the response to pesticides. The use of different sized litter bags presents differences in their relationships with the surrounding soil and in the degree of compression of the material contained in them. These methods almost exclusively concentrate on leaf material and so ignore the considerable root biomass (up to fifty percent of the total plant material) whose degradation may be affected by a pesticide. One alternative which eliminates some, if not all, of these differences is to use a standard 'model' substrate such as pure cotton (cellulose) cloth. This approach has been used widely by soil microbiologists and has been applied by Wingfield (1980) to effects of herbicides. He used a standard test cloth (Shirley test cloth, The Shirley Institute, Didsbury, Manchester, U.K.) which overcomes problems of variability of weave and cloth type inherent with other materials. Standard sized strips of this cloth may be mounted on microscope slides to give rigidity and facilitate burial in soil, either in the laboratory or in the field. Assessment of degradation can be made on the basis of weight loss, though the cloth is designed for measurement of tensile strength. As with all soil burial techniques, a major problem is variability of results. However, the technique is so simple and economic of labour and resources, that replication levels can be very high.

A similar technique has been developed for studying effects on nitrogen mineralization (Bebb and Greaves, 1983). This involves burial of pieces of colour transparency film, mounted in plastic slide mounts. The protein emulsion layers produce areas of different colours on degradation which can be assessed by measuring weight loss, nitrogen loss or as area affected using an image analysing computer. A similar approach can be taken using black and white film, though in this case the emulsion layer is so thin that weight loss becomes an unreliable indicator. Again, variability is high but can be overcome by increased replication. Sensitivity, as found by Bebb and Greaves (1983) is certainly adequate to detect the degree of change that is deemed to be significant.

One final comment concerns the use of test papers. For many years the use of pH indicator paper has been one accepted way of measuring soil acidity and predicting the need for liming. Is it any different to use more recently developed indicator strips for nitrate, nitrite or ammonia determination? Admittedly, the degree of precision they offer is low but it is more than adequate to detect that large changes have occurred and so indicate the need for complex, expensive analysis.

CONCLUSIONS

It has been argued that it is perfectly feasible to introduce a simple, cheap, preliminary tier of tests to present protocols. This preliminary examination would, with acceptable sensitivity, identify those chemicals with the potential to cause harm to soil microflora and which would, therefore, need further testing. By the same token, it would exempt many chemicals from this need and so reduce the costs of development.

This argument does not imply that this preliminary scheme, or further detailed testing by present methods, is satisfactory as an absolute predictor of likely effects in the soil environment. Indeed, continued detailed research into this area is essential. It is to be hoped that this occurs and, more particularly, that it concentrates on the development of more realistic and valid methods than are available today.

REFERENCES

Beeb, J.M. and Greaves, M.P., 1983, British Journal of Photography, 22, 570.

Cook, K and Greaves, M.P., 1986, in Pesticide Effects on Soil Microflora edited by L. Somerville and M.P. Greaves, (London, Taylor and Francis Ltd.) p 250

Cooper, S.L., Wingfield, G.I., Lawley, R. and Greaves, M.P., 1978, Weed Research, 18, 105.

Greaves, M.P., Poole, N.J., Domsch, K.H., Jagnow, G. and Verstraete, W., 1980, Recommended Tests for Assessing the Side-Effects of Pesticides on the Soil Microflora, Technical Report, Agricultural Research Council, Weed Research Organization, No. 59.

Wingfield, G.I., 1980, Bulletin of Environmental Contamination and Toxicology, 24, 473.

CHAPTER 14

PESTICIDE SIDE-EFFECTS:
REGULATIONS IN THE NETHERLANDS

A. M. van Doorn

INTRODUCTION

The Committee for the Registration of Pesticides in the Netherlands includes several groups that advise on the registration of pesticides and whose members represent various ministries. One group deals with pesticides, another group with disinfectants, a third group with wood preservatives, and so on. In 1975 a subgroup of scientists was set up to focus on the environmental fate and effects of pesticides.

The first task of this subgroup was to formulate the requirements for data on the behaviour of a pesticide and its transformation products in soil, water and air as well as on its toxicity to individual non-target organisms in the environment and the biomass in soil.

In 1975 the first edition of the registration requirements was published. It is currently being revised, in the light of ten year's experience of evaluating the environmental data on pesticides submitted by the agrochemical industry.

During the last ten years the subgroup has spent much time on its second task, evaluating the many environmental data produced by the agrochemical industry. The aim of the subgroup is to collect all data on the individual pesticides and to provide the decision-makers with a report referring to each pesticide's possible risks to the environment (e.g. accumulation in soil and water in relation to persistence; the degree to which the physico-chemical properties, rate and time of application affect the risk of the compound and metabolites leaching to the groundwater; the risk to water organisms in the aquatic ecosystem).

In the process of evaluating the data on the behaviour of pesticides, most attention is paid to rate and route of degradation, leaching and the persistence and accumulation of residues.

When formulating the requirements in 1975, considerable effort was spent in bringing them into line with those formulated by the Environmental Protection Agency in the USA. It is worth mentioning that in the USA a different procedure is followed to determine the requirements needed to judge whether a pesticide is admissible. In the USA all interested parties are invited to submit written comments on the proposed regulation requirements to the EPA: this generates an ongoing debate about the pragmatism of those requirements.

In the Netherlands, until recently, the only data required on a pesticide's side-effect on non-target soil organisms were those on effects on soil respiration and nitrogen conversion. This was in accordance with successive workshops on "side-effects of pesticides on soil microflora", which recommended that data on the following side-effects of pesticides be mandatory:-

a) the effect on soil respiration, measured from carbon dioxide generation or oxygen consumption and

b) the effect on nitrogen mineralization and nitrification.

These data had to be submitted unless it could be demonstrated that the substance would not penetrate the soil (this is applicable to only a few compounds). The measurements had to have been taken in at least two soil types, whose characteristics were stipulated in the guidelines. If the pesticide was intended to be applied in an area with a soil type that differed from the stipulated types, then the pesticide had to be tested on that soil type. Finally, it was decided that if the effects on soil respiration and/or nitrogen conversion were of a long-term nature, more detailed information might be required, especially on the effects on other microbial or enzymatic processes such as urease and/or phosphatase activity in the soil, and on nitrogen conversion under field conditions. Until recently, no other requirements (e.g. on toxicity to soil organisms) were stipulated: this in contrast with the requirements for ascertaining the effects of pesticides on water organisms such as algae, invertebrates (Daphnia) and vertebrates (a fish species).

In 1984, the Dutch registration requirements of 1975 were reviewed, partly in response to comments from the agrochemical industry, partly because of our own experience in evaluating environmental data, but especially because of the realization that the only data we need are those that are directly relevant to the evaluation and that contribute substantially to the process of designating a pesticide as admissible.

To give an example: the great concern of the agrochemical industry to perform a study on degradation of a pesticide in surface water only in a pure water system has led now to a requirement for the degradation in a water/sediment system, which is much more realistic.

Another example: during the last ten years it has become apparent that the mandatory data on soil respiration and nitrification have not played any role in the evaluation of the environmental impact of pesticides. It has been found that any effect on soil respiration is soon put right.

In some cases, however, especially in the application of soil treatment compounds (soil fumigants and granulates), a significant inhibition of nitrification was found. In these cases the agricultural impact appeared to be of more importance than the repercussions to the environment.

Supported by the conclusions of the 1984 Versailles conference and by the recent edition of the EPA guidelines which no longer require data on soil respiration and nitrification, the subgroup of Dutch scientists proposed scrapping these requirements, except the requirement for nitrification in the case of compounds that are incorporated into the soil. This does not mean, however, that the Dutch authorities are no longer interested in effects on soil organisms: they are, but only if the data are reproducible. The main problem with assessing the effects of a pesticide on soil respiration and nitrification is that nowhere is it spelt out precisely what is relevant information on the side-effects on soil microflora and how this information should be weighed when judging a pesticide's admissability.

As to what is relevant information; more should be known about the effects of a pesticide on specific soil processes or specific organisms that are very important to soil fertility and quality. Except for the case of nitrogen fixation, this question has not yet been solved because of the lack of relevant methods. For example, for many years excellent research has been done on the influence of pesticides on the population of springtails and mites in the soil, because of the important role these organisms play in maintaining soil structure. With the exception of the long-term influences of very persistent compounds and DDT, the influence of pesticide application appeared to be neither greater nor longer-lasting that that caused by drought, soil moisture, or cultivation measures. In all cases the recovery time of the disturbed populations appeared to be of the same order! The case of soil respiration and nitrification is similar. The Dutch authorities have not introduced a test to ascertain the effects of a pesticide on springtails and mites: however, in 1984 they did introduce a test on toxicity for earthworms after intensive discussions had indicated that the earthworm can be considered not only as an indicator organism, but also as an organism of special significance for soil structure, especially in meadows.

As to how we should weight information on the effects of a pesticide on soil microflora, the problems that arise are how stable is soil quality (for example, is there a difference in this respect between arable land, forests and nature areas).

In summary, in the Netherlands the information required on the side-effects of pesticides on soil microflora is restricted to the effect of soil-treatment compounds on nitrification, and the acute toxicity on earthworms not because the Dutch authorities believe that side-effects on the structure and function of the soil ecosystem are of minor importance but because of the lack of appropriate methods that provide relevant data for evaluating a pesticide's effects in the soil.

Here it is useful to compare the Dutch case with the Environmental Protection Agency's history of formulating guidelines for registering pesticides in the USA.

The 1975 version of the EPA's proposed guidelines for soil metabolism studies stipulated that the data required had to provide:-

a) information on the rate, type and degree of degradation of the parent pesticide and its metabolites,

b) information on the pesticide's persistence in soil,

c) information on the effects of pesticides on microorganisms.

The argument behind these requirements was that: pesticides reaching the soil could significantly reduce or eliminate populations of non-target soil organisms and thereby alter the biochemical characteristics of the soil, possibly resulting in increased persistence of a pesticide or harming relevant enzyme systems of natural soil populations

d) information on effects of microorganisms on pesticides

Here, the argument was that in general, biodegradability is a major factor affecting the persistence of a pesticide. Where biodegradation is of great importance, a qualitative and quantitative assessment of the role of microorganisms in degradation is required.

The 1980 edition of the EPA proposals contained requirements for data on the effects of microbes on pesticides and of pesticides on microbes. On the one hand the argument was that microbes, one of the most important groups of organisms in the soil involved in the biochemical transformation of pesticides in soil, may affect the rates of degradation. Microbes may also function in maintaining soil fertility. On the other hand, the effects of pesticides on the function of microbes is not always easy to discern. Pesticides may inhibit those populations that degrade other pesticides and so increased persistence of pesticides applied in tandem or in sequence may result. There are so many possible pesticide combinations, that it is impossible to test each of them for antagonistic effects to all microbes.

The interaction between pesticides and microbes may affect the availability of the pesticide to non-target organisms and/or the accumulation of the pesticide in the food web, or may result in the loss of usable land. Cases are known of the crumb structure of soil being decreased because of the elimination of fungi by pesticides, with a consequent reduction in the water-holding capacity.

To cover all these questions the EPA proposed protocols to ascertain the impact of pesticides on the function of resident populations of soil microorganisms involved in nitrification, nitrogen fixation and degradation of complex molecules such as cellulose, starches, pectin and other materials.

Representative organisms to be tested with a corresponding substrate were listed for each of the following functional categories. Two or more organisms must be tested:-

1) free-living nitrogen fixers (Azotobacter, Clostridium, Arthrobacter)

2) nitrifiers (Nitrosomonas, Nitrobacter)

3) cellulose degraders (Cellomonas, Cytophaga)

4) starch degraders (Chaetomium, Streptomyces)

5) pectin degraders (Flavobacterium, Bacillus, Pseudomonas, Penicillium, Trichoderma, Aspergillus, Fusarium)

Finally, after intensive discussion, the 1982 edition of the EPA guidelines no longer contains requirements referring to the effects of pesticides on microbes and microbes on pesticides. This is because the EPA prefers a tier system of testing, starting for example with requirements for rate and route of degradation. If the evaluation indicates a certain risk for the environment, more field monitoring studies may provide the information necessary to evaluate the significance of the risk to non-target organisms.

Pending the development of properly designed and validated protocols from which useful regulatory conclusions can be drawn on the role of microbes in the overall environmental fate of pesticides, the EPA no longer stipulates data on the effects of microbes on pesticides, or on the effect of pesticides on microbes.

If the conditions for proper protocols are fulfilled, the EPA will probably include all such microbial studies in a separate subdivision of the guidelines.

At present, the current Dutch situation regarding the influence of pesticides on the soil microflora is the same as in the USA. There is inadequate knowledge of what a soil ecosystem is and which processes and organisms are of major importance in maintaining soil quality.

Not surprisingly, in the Netherlands today the priority in research is for research on soil ecosystems and on the influence of human manipulations on these systems.

CHAPTER 15

LIMITATIONS TO PRESENT METHODS

M. P. Greaves & L. Somerville

INTRODUCTION

One of the most important functions of the international workshops on side-effects of pesticides on the soil microflora has been to identify the limitations on present methods.

In the following we have identified the relevant conclusions from the foregoing chapters:

Limitations to present methods

1. The soils selected for use, especially the so-called 'standard' soils often contain too low a microbial biomass.

2. Storage of soil prior to experimentation is often unavoidable. Unless great care is taken to ensure and maintain proper storage conditions, considerable loss of biomass can occur. Shifts in microbial dominance patterns, even in good storage conditions, cannot be avoided.

3. Under field conditions plant exudates and detritus make major contributions to the maintenance of the soil microflora. In laboratory experiments these contributions are absent.

4. Although the recommended tests give reasonably reproducible results within one laboratory, there can be very high variation between laboratories for the results in the same test with the same pesticide.

5. There is a substantial lack of unequivocal evidence to show that the data from laboratory experiments reliably effect the situation in the field.

6. The data obtained from the laboratory experiments can be biased by the choice of sampling techniques and by experimental design.

7. There is often no attempt to evaluate the microbial side-effect data in the light of the degradation of the pesticide under test. Thus, side-effects may be found and ascribed to a pesticide long after it has been effectively degraded.

8. Measurement of long-term soil respiration, especially in the absence of added nutrient, is a relatively insensitive means of detecting side-effects due to pesticides.

9. The test of effects on ammonification is thought by some to be inadequate for assessing side-effects.

10. Prediction of field effects from laboratory data is not possible.

11. Measurements of cumulative long-term respiration can, in some cases, show increases in CO_2 evolution which may be due to a lethal effect of added pesticide on some of the microflora providing readily available substrate for the survivors. This increase is not yet necessarily regarded as a harmful side-effect. There may be similar increases in nitrogen mineralization in certain circumstances.

The consequences of such limitations are perhaps best summarized in Chapter 14 by van Doorn. He states (p 12) "At present, the current Dutch situation regarding the influence of pesticides on the soil microflora is the same as in the USA. There is adequate knowledge of what a soil ecosystem is and which processes and organisms are of major importance in maintaining soil quality.

Not surprisingly, in the Netherlands today the priority in research is for research on soil ecosystems and on the influence of human manipulations on these systems".

It is summarized in another way in the current recommendations of the last Workshop (see Appendix B). "These recommended tests are only seen as those most appropriate at present". <u>More research is essential</u>.

F.A.O. EUROPEAN COOPERATIVE RESEARCH NETWORK ON PESTICIDES

H. Schuepp

The "FAO-European Cooperative Research Network on Pesticides with Special Reference to their Impact on the Environment" (abbrev. "Network on Pesticides") seeks to bring together specialists competent in specific scientific topics and generalists trying to work out a synthesis. Not only have interinstitutional links been established between governmental authorities responsible for agriculture, environmental protection or public health but also valuable contacts have been achieved between universities, research stations and private industry.

Within the "FAO Network on Pesticides" scientific aspects are being discussed, and protocols are being worked out, concerning side effects of pesticides on non-target organisms.

With the support of the FAO, the recommendations for meaningful testing of side effects of pesticides on soil microflora could further be re-appraised and amended. Clear and thoroughly discussed <u>recommendations</u> and <u>guidelines</u> are needed by governmental authorities and for international standardization of the requirements.

The "Network on Pesticides" tends not only to be international but also interdisciplinary and interinstitutional. It tries to establish contacts and to promote exchange of information among the various International Organizations active in this field as well as to enhance public relations.

The Coordination Board elected at the Consultation in Versailles in 1984 is comprised of the Coordinator of the Network (Chairman of the Board) and the Officers-in-Charge of the Liaison Centre of the Sub-Networks. Each member of the board covers a distinct research area thus promoting interdisciplinary coordination and cooperation.

The Coordination Board began its responsibility at the meeting in Wädenswil, 27–28 June 1985, when future activities were extensively discussed.

ORGANISATION OF THE NETWORK: COORDINATION BOARD

Chairman of the Coordination Board, Liaison Centre for Effects of Pesticides on Non-target Microorganisms (Dr. H. Schüepp, Swiss Federal Research Station, CH–8820 Wädenswil, Switzerland).

Liaison Centre for Effects of Pesticides on Water Ecosystems, Microbial Processes in Soil and Nutrient Cycling (Dr. M.P. Greaves, Head of Environment and Vegetation Management Group, AFRC, Institute of Arable Crop Research, Long Ashton Research Station, Bristol, BS18 9AF, England).

Liaison Centre for Behaviour of Pesticides in Soil, Degradation, Adsorption and Mobility (Dr. M. Hascoet, Directeur, Department of Phytopharmacy, INRA, National Institute for Agricultural Research, F–78000 Versailles, France).

Liaison Centre for Effects of Pesticides on Warmblooded Animals, Toxicology, Behaviour, Ecosystems Changes (Dr. D. Osborn, Institute for Terrestrial Ecology, Monks Wood Experimental Station, Huntingdon, PE17 2LS, England).

Liaison Centre for Effects of Pesticides on Soil Fauna, Root Dynamics and Rhizosphere Interactions (Prof. Dr. E. Steen, Department of Ecological and Environmental Research, Swedish University of Agricultural Sciences, S–75007 Uppsala, Sweden).

Two Working Groups–"Inorganic Bromide in Soils" and "Root Dynamics and Aspects of Rhizosphere Biology"– have been formed. Within these groups it is hoped to establish Joint Research Projects. Dr. Hascoet has initiated the working group on Inorganic Bromide in Soils. Topics to be investigated jointly or in ring-tests will be such things as leaching of bromide to groundwater, transfer of bromide from soil to plants and alternatives to methyl bromide. A working group on Root Dynamics and Aspects of Rhizosphere Biology will be coordinated jointly by E. Steen, M.P. Greaves, and H. Schüepp. It is hoped that interdisciplinary investigations in several Institutes might be realized based upon similar experimental setups or long-term field experiments.

A Network-Bulletin in the form of a Newsletter is published. It has a free circulation to Representatives in all participating countries. The aim of the bulletin is to pass information about the activities of the network and related organisations. It includes, invited articles on subjects of current concern, identifies areas where research is needed or where it might be reduced. The bulletin is also sent to other regions of the world and to other international organisations. Copies of the Bulletin and further information concerning the Network can be obtained from Dr. H. Schüepp.

RECOMMENDED LABORATORY TESTS FOR ASSESSING THE SIDE EFFECTS OF PESTICIDES ON SOIL MICROFLORA.

PROCEEDINGS OF THE THIRD INTERNATIONAL WORKSHOP, CAMBRIDGE, SEPTEMBER 1985

<u>Organizing and Editorial Committee</u>

L. Somerville   Schering Agrochemicals Limited, Chesterford Park Research Station, Saffron Walden, Essex CB10 1XL, UK.

M.P. Greaves   Long Ashton Research Station, Weed Research Division, Long Ashton, Bristol BS18 9AF, UK.

K.H. Domsch   Institut für Bodenbiologie, Bundesforschungsanstalt für Landwirtschaft, D 3300 Braunschweig, F.R. of Germany.

W. Verstraete   Lab. Microbial Ecology, University of Gent, Coupure L 653, B 9000, Gent, Belgium.

N.J. Poole   ICI Plant Protection Division, Jealott's Hill Research Station, Bracknell, Berkshire RG12 6EY, UK.

H. van Dijk   Institute for Soil Fertility, PO Box 3003, 9750 RA Haren Gn, The Netherlands.

J.P.E. Anderson   Bayer AG, PF-F/CE, Institut für Ökobiologie, Zentrum Landwirtschaft, Monheim, D 5090 Leverkusen, F.R. of Germany.

INTRODUCTION

During the period 1973-77 four Symposia, dealing with pesticide side-effects on non-target soil micro-organisms, were organized jointly by the Biologische Bundesanstalt and the Bundesforschungsanstalt für Landwirtschaft, Braunschweig, F. R. of Germany. As a result of these meetings it was decided to extend the participation at subsequent meetings. Consequently an International Workshop was held at Braunschweig in 1978 followed by a second Workshop in England in 1979. The aims of these meetings were to discuss present knowledge of the means of testing side-effects of pesticides on the soil microflora and to agree on recommendations for meaningful tests, suitable for application as registration requirements.

Following the 1979 Workshop, agreed recommendations for testing side-effects of pesticides on soil microflora were published and distributed widely to relevant individuals and organizations (Greaves et al. 1980). Consequently, some of these tests, viz. the soil respiration and nitrogen transformation tests, were adopted by some regulatory authorities and considerable experience has been gained in their application and interpretation. Following more than 5 years use of the recommendations it was decided to hold a further Workshop to consider their current status and to discuss the need for revision in the light of experience.

This document summarizes the proceedings of this latest Workshop and presents the revised recommendations agreed by the delegates present. A list of these delegates is appended. They represent government research institutes, universities and industry and are involved in studying side-effects of pesticides on the soil microflora, either for registration purposes or for agricultural research.

As in 1979, the unanimous decision of the 1985 Workshop was that the recommendations should be circulated as widely as possible to persons involved with, directly or indirectly, or interested in the impact of pesticides on the soil microflora. The hope was expressed that, in this way, the recommendations will reach those responsible for framing and operating pesticide registration schemes.

At the current stage of knowledge there are some reservations about the relevance of existing tests to the environment. However, there must be some safeguards, however crude, regarding the effects of artificial chemicals on soil fertility.

In spite of the lack of totally acceptable criteria,

we detail below those tests with which we have sufficient experience to provide guidance on their use. These, in the absence of more suitable tests, may be used, under appropriate circumstances, for registration purposes.

1.   PRE-REGISTRATION HAZARD EVALUATION AND PREDICTION

The risks to the environment from a pesticide or its formulated product are dependent on many factors. For example, its toxic properties, persistence and mobility in the environment, the amount applied, the formulation, method and time of application and, particularly, frequency of use are all important.

Some pesticide effects on the environment may be too complex, subtle or delayed to be detected by ordinary testing in the laboratory or the field. In any case, it is obviously impossible to cover in such trials the infinite variety of conditions under which the pesticide may be used in practice. Nevertheless, experience has shown that in many instances, predictions can be made of probable environmental effects of a compound from consideration of certain basic information.

Data have to be obtained prior to registration to allow a reasoned judgement to be made of the environmental behaviour of the product, when applied according to the recommendations for use. Such data are essentially predictive and are intended to describe those characteristics of the product relevant to the environment. They should be sufficiently comprehensive to enable the authority to make a reasonable judgement of the environmental behaviour of the product for the uses proposed. They do not seek to give data on the actual behaviour of the product in all the many environments in which it will be applied or which might be reached by the pesticide. The actual test programme has to be decided according to the characteristics and conditions of use of the product.

2.   PRIMARY DATA ESSENTIAL FOR PREDICTING ENVIRONMENTAL HAZARDS

The information given below on the use pattern is an essential element for the estimation of the expected environmental concentration and the probable sites of deposition.

Properties of the pesticide:
Physico-chemical properties of pesticide

Biological activity on target species
Metabolism and residue studies, including persistence and
mobility
Mammalian toxicology
Toxicological data on other species

Use pattern:
Formulations
Methods of application
Site of application
Time of application
The amount applied
Scale of use
Climate and geographical locality.

2.1 ASSESSING THE NEED FOR TESTING SIDE-EFFECTS

If a pesticide is applied directly to the soil, and
if a potential hazard is predicted from the primary data,
the relevant studies selected from those recommended in
Section 3 should be done.
If a pesticide is not applied directly to the soil,
but it is considered likely that it might reach soil,
relevant studies of effects on soil microflora should be
considered.

3. RECOMMENDED LABORATORY TESTS

3.1 TEST CONDITIONS

Selection of soils

At least two widely occurring agricultural soils,
representative of soils on which the pesticide is likely
to be applied, should be chosen. They should represent
conditions where the soil microflora may be under (a)
relatively high stress from the pesticide, (b) relatively
low stress from the pesticide. (In many cases these will
be a sandy soil low in organic matter, and a loamy soil).
The stress from the pesticide can be evaluated using the
physical and chemical properties of the compound and from
the degradation rate of the pesticide and its metabolites
in various soils. Where appropriate, one or more of the
soils recommended by OECD may be considered (Anon.,
1981). The soil should have a usual cropping pattern and
preferably have received no pesticide, or only pesticides
which are known not to affect microbial processes, for
five years. Soil contents of total carbon, clay, silt,

sand, pH and time of sampling should be stated. In addition, an appropriate measurement of the biological status of the test soils should be stated (e.g. biomass).

It is suggested that, wherever possible, pesticide degradation studies should be done on the same soils as those used for side-effect assessment.

Soil collection and treatment

When sampled, the soil should, wherever possible, be at a moisture content suitable for sieving. The top 10 cm only should be used and the vegetation, soil animals (macro-fauna) and stones should be removed. The soil should be passed through a 2 mm sieve. If too wet to sieve, it should be dried by spreading but must never be air dry (less than c. 30% water holding capacity) as this adversely affects the microbial biomass. The soil should be thoroughly mixed before use.

Soil storage

The object of the subsequent tests is to investigate the effects of a pesticide on the soil microflora. It is, therefore, desirable that the soil used should be as fresh as possible. If storage is necessary, loss of microbial biomass can be minimized by keeping the soils at 2°-4°C in such a way as allows adequate access of air. Drying, water-logging or freezing of the soil must be avoided at all stages of storage and/or treatment. The soil should not be stored for more than 10 weeks and before use should be kept at 20±2°C for a minimum of 2 to a maximum of 14 days to allow for equilibration.

Pesticide dosage

1. Recommended field rate (kg ai/ha) expressed as mg/kg (dry weight) soil, assuming uniform distribution in the top 5 cm of the soil. Fumigants should be used at a dose corresponding to the recommended application, taking the actual time of exposure into consideration.

2. Ten times the recommended field rate. Fumigants at 5 times the recommended concentration, taking the actual time of exposure into consideration.

    If the pesticide is applied in a solvent, the appropriate solvent control should be used in addition to the untreated control.

3.2 SOIL RESPIRATION TESTS

The most suitable method is the determination by continuous or semi-continuous monitoring, of the evolution of CO_2 from soil aliquots (100 g) kept in the dark at 20 $\pm$2°C. The soil moisture content should be pF 2.0-2.5 or kept in the range between 40 and 60% of the water holding capacity. It should never exceed the upper plastic limit! The tests should run for a minimum period of 30 days, the actual duration being dependent on the reaction of the soil respiration to the compound and/or the added organic amendment.

The recommended test procedure is as follows:

TIER 1.   Use the most sensitive soil of those selected (Section 3.1) (usually less than 1%C) and amend with 0.5% w/w plant or horn meal (C:N ratio less than 16, milled to pass a 0.5 mm sieve).

The treatments to be applied are:

1. No further treatment
2. The highest recommended pesticide field rate
3. Ten times the highest recommended field rate

TIER 2.   Use the most sensitive soil as in TIER 1 but do not amend with plant or horn meal.

The treatments to be applied are as for TIER 1.

TIER 3.   Use a second soil (usually greater than 1.5% C) both amended and unamended with 0.5% w/w plant meal.

The treatments to be applied to both these are as for TIER 1.

Should the pesticide treatments in TIER 1 cause no critical* deviation from control results no further tests are required. If such a critical deviation does occur proceed to TIER 2. Progress from TIER 2 to TIER 3 is based on the same criteria.

* For a definition of 'critical' see Section 4

3.3 LITTERBAG TEST

This test should only be required when 'critical'side-effects are detected with the plant meal amended respiration test.

Wheat straw (haulms cut in c. 2 cm internodal pieces) is recommended, though other plant materials may be substituted depending on the intended use of the pesticide (e.g. apple leaves for pesticides generally used in orchards). Nylon bags (10 x 10 cm, 2 mm mesh) are filled with 2 g (dry matter) of straw. The straw should have been treated (by spraying or dipping) with a pesticide dosage equivalent to that normally applied to 100 cm^2 soil area. To reduce variability and to prevent plant roots penetrating the bags, cylinders (20 cm diam., 10 to 20 cm length) are sunk into the soil around the bags, which are buried at a depth of 5 cm. There should be at least 6 replicate bags for each sample date. The experimental field site should have the main crops for which the pesticide is intended. Samples should be taken at the middle and at the end of the vegetative growth period. Decomposition is measured gravimetrically.

3.4 NITROGEN TRANSFORMATION

AMMONIFICATION, the release of ammonium from organic matter, is of agricultural importance. However, it is performed by a wide variety of soil micro-organisms and is therefore relatively insensitive to perturbation. Tests of pesticide effects on this process are not essential, but data can be obtained conveniently by using an organic substrate instead of $(NH_4)_2SO_4$ in the nitrification test.

In contrast, NITRIFICATION, the oxidation of ammonium to nitrite and nitrate, is carried out by fewer species of micro-organism and is of considerable ecological and agricultural importance. Effects of the pesticide on nitrification should therefore be investigated.

Nitrification can be studied using soil amended either with $(NH_4)_2SO_4$ or with organic matter, such as plant or horn meal as described for the respiration test. For this purpose $(NH_4)_2SO_4$ or organic matter equivalent to ca. 100 mg N/kg soil are added to the soil and the disappearance of NH_4^+ and the appearance of NO_3^- is followed. The appearance of NO_2^- need only be checked when NH_4^+ disappears and NO_3^- does not appear. In all cases, it is essential that it is first proved that the soil is capable of nitrification and that the organic matter amendment is suitable for ammonification and nitrification studies.

The recommended test procedure is exactly as described in Section 3.2.

The study of nitrification has to be continued until either the added substrate has been converted or until an equilibrium has been reached between ammonification from soil organic matter and nitrification.

Care must be taken, especially with clay soils, to avoid anoxia which might cause denitrification.

3.5 NITROGEN FIXATION

Tests should be restricted to situations where symbiotic nitrogen-fixation may be affected by the pesticide, i.e. where pesticides are used directly on legumes, where legumes are a normal constituent of the crop rotation system, or where mixed leguminous/non-leguminous cropping is practised. The choice of the test legume should depend on the 'use pattern' of that particular pesticide.

The assessment method should recognise the unique symbiotic relationship between the host plant and Rhizobium, and hence should determine in one test both plant and bacterial response to the pesticide. It therefore will be required to measure plant yield and nitrogen fixed. These experiments will be conducted in pots containing a soil suitable for growth of the plant. The plant, seeds or soil should be inoculated with Rhizobium only if no suitable bacteria are present naturally in the soil. When required, the plants will receive a mineral fertilizer which excludes nitrogen.

If the pesticide causes a 'critical' effect in these tests, then field data should be provided.

For pesticides to be used in rice culture, the effects of the pesticide on the growth and nitrogen-fixing ability of a Cyanobacterium (e.g. Anabaena) and on the Azolla/Anabaena symbiosis should be determined.

Present knowledge indicates that non-symbiotic nitrogen fixation in soil and in the rhizosphere of agricultural crops does not warrant regulatory studies at this stage. We would, however, recommend this area for further research funding.

4. INTERPRETATION AND EVALUATION OF DATA

The problem of deciding the relative importance of each observed effect a chemical may have is fundamental to side-effect testing. It has to be decided how to interpret or evaluate the importance of the sum of effects, both inhibitory or stimulatory, which could occur when investigating multiple microbial activities.

The meeting agreed that there is an urgent need to develop means of objectively assessing the data obtained from recommended tests of pesticide side-effects. An outline of one system being developed (Domsch et al., 1983) is given here.

4.1 TYPES OF MICROBIAL RESPONSES

Chemicals with biocidal properties can produce a limited number of inhibitory as well as stimulatory responses in reactive biological systems (Figure 1). The effects themselves may be reversible or irreversible within the monitoring period. The four resulting types cover most of the responses thus far observed. Combined

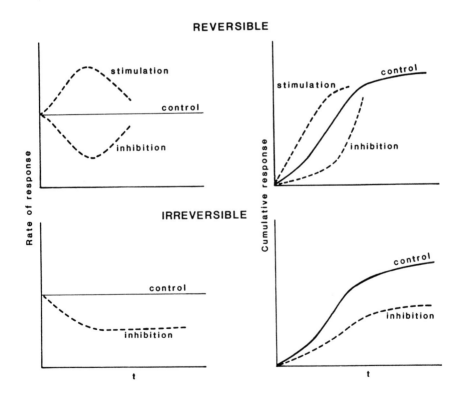

Figure 1   Principal types of reversible and irreversible responses

responses may occur, in which case the smaller effect (less impact, shorter duration) should receive less attention.

It is evident that any monitoring system which would allow such an uncomplicated recognition of reaction types requires high quality data. Since a chemical which is incorporated into the soil reacts in a dynamic system which has a multitude of ways and means for repair and recovery, the time factor has a key position. Side-effect data, without information on the time-dependent changes of the microbial response, are only of limited use for evaluation purposes.

The reaction types indicating reversible effects may be analysed more specifically in order to recognise their quantitative aspects. There are two main criteria which can be applied to all reversible negative responses:

(a) the approximate amplitude of the maximal stimulation or inhibition, and

(b) duration until recovery (Figure 2). It should be mentioned here that, from an ecological point of view, irrespective of the magnitude of stimulation or inhibition, the delay in the re-establishment of an affected microbial structure or function ranks higher in the hierarchy of evaluation criteria.

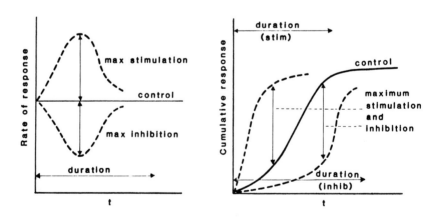

Figure 2   Criteria used for the evaluation of reversible side-effects

4.2 INTERPRETATION OF RESULTS

Reversible effects

The recognition of response types is a pre-requisite for further evaluation, but does not allow a decision to be made on which effects might be ecologically insignificant, tolerable or even critical. In order to approach this problem, it is proposed that in situations where the application of pesticides is justified from an agricultural point of view, the magnitude of response to the man-made chemical stress should be compared with that of naturally occurring stress situations. From studies in microbial ecology a wealth of information is available on the consequences of naturally occurring 'catastrophes' in soil microhabitats: water availability is frequently reduced to critical levels or soil is flooded and gas exchange is limited; the temperature often deviates from optimum conditions or nutrients are lacking and cells are eliminated by soil animal grazing, micro-parasites or other adverse conditions. For quantitative data see Domsch et al. (1983).

From ecological data, it can be concluded that depressions of 50% or more, frequently occur under natural conditions. Soil meteorological data prove that wide variations in microenvironmental conditions are normal events. If these statements can be accepted, depressions due to man-made chemical stress must not be overestimated, as long as they are of a magnitude comparable to those found in natural situations.

Likewise, the determination of the time required for the restitution of normal microbial populations or functions (i.e. duration until recovery) after the end of natural stress conditions is required as an ecological 'yardstick'. A key parameter is the doubling-time of microbial cells. Under field conditions doubling-times are considerably longer than under optimum laboratory conditions. A realistic estimate is probably a doubling-time of about 10 days.

If this basic consideration can be accepted, delay (recovery) periods of 30 days following an approximately 90% depression are still a normal and natural phenomenon. Applied to the situation of chemical stress, a delay of 30 days should be regarded as being of 'no ecological significance'. Delays of 31 - 60 days can still be 'tolerable' while those of more than 60 days indicate 'critical' situations.

Irreversible effects

The side-effects which in laboratory experiments are apparently irreversible require separate consideration from the two above-mentioned response types. These side-effects are characterised by deficits of populations or functions when pesticide-treated soils are compared with non-treated soils at the end of monitoring periods which have been long enough to allow for recovery. The deficits may be due to (a) the persistence of the toxic agent or (b) the inability of a population to recover.

Response to adverse effects

In general, adverse reversible or irreversible effects are recognized from high (50%) or long-lasting (60 days) deficits. In these instances, two steps should be followed:

(i)   Confirmation of test data by repeating the laboratory tests using different environmental conditions.
(ii) If the test data are confirmed in (i), field tests should be done. In the case of field testing, the methods recommended for use in the laboratory may be appropriate.

5. **TESTS WHICH ARE UNSUITABLE FOR MEASURING SIDE-EFFECTS OF PESTICIDES ON SOIL MICROFLORA**

5.1 **AXENIC CULTURES**

The meeting considered that, though useful in certain research studies, plate counts and pure culture experiments are unsuitable for regulatory studies because:

Isolated organisms may be metabolically atypical of their form in soil. Furthermore, they can change progressively during storage.

They are normally stimulated to artificially high metabolic rates by growth in nutrient rich laboratory media.

They are removed from their normal ecological associations.

Interpretation of results is difficult and extrapolation to field situations impossible.

216

Thus, they are not able to reflect, in any meaningful way, the side-effects pesticides may have. Side-effects are best tested for by observing microbial processes in soil.

5.2 SOIL ENZYMES

The consensus of the meeting was that, with the present state of the art, soil enzyme activities would be of little value when monitoring side-effects of pesticides on the microflora. The principal reasons are as follows:

1. The total enzymic capacity of a soil is made up of various fractions (i.e. proliferating microbes, substrate turnover in dead cells and debris, enzymes associated with colloidal humic-matter etc.) and it is extremely difficult to quantify the contribution of each to the catalysis of a particular substrate.

2. Current research specifically identified the term 'soil enzymes' with the humic-matter and, although this is believed to play an important role in substrate turnover, there is no universally agreed methodology. Almost any result can be achieved by varying assay conditions (temperature, pH, substrate concentrations).

Phosphatase and dehydrogenase have been suggested as useful enzymes for monitoring pesticide side-effects. Even if the previously stated drawbacks were resolved, it is believed that these two enzymes are unsuitable. Phosphatase is a collection of enzymes, usually measured by using an artificial substrate (e.g. p-nitrophenyl phosphate), and whose activity bears little relation to total phosphate availability in soils. Dehydrogenases reflect a broad range of microbial oxidative activities, do not accumulate to any extent (based on the definition above, they are not soil enzymes) and yet they do not consistently correlate to microbial numbers, CO_2 evolution or O_2 consumption. Additionally, activity may be dependent upon the nature and concentration of amended C-substrates and alternative electron acceptors.

6. FUTURE DEVELOPMENTS

The meeting agreed that considerably more research is required to devise laboratory tests that can be used predictively with some confidence.

217

Applied research should be directed at ways of improving our ability to identify, quantify and evaluate potential harmful effects of pesticides to soil microbial activities, which are essential to the maintenance of soil fertility. World agriculture is dependent on the continuing supply of effective, safe chemicals for pest control. Any requirement which adds significantly to development costs, without making a valuable contribution to environmental safety, must be evaluated most critically.

Basic research is the only way in which the information essential to the applied scientists can be obtained. As such it must take considerable priority in competition for funding. It is essential that we understand microbial behaviour in soils more thoroughly. In particular, the interactions between pesticide, soil, crop plant and micro-organisms require continued exhaustive research.

Obviously, the choice of important specific areas of research depends, to a great extent, on individual bias. However, the following topics seem to warrant special attention with regard to the effect of individual pesticides and of combinations and sequences of pesticides.

1. Rhizosphere studies - in particular nodulation of economically important legumes and mycorrhizal associations.
2. Soil-borne plant pathogens including microbial antagonism.
3. Microbial aspects of soil structure.
4. Microbial aspects of nutrient cycling, other than nitrification.
5. Interrelationships between field and laboratory data.
6. Interactions between microflora and microfauna.
7. Additive stress arising from use of mixtures and sequences of chemicals and the simultaneous occurrence of natural and chemical stress situations.

These subjects can be studied effectively only if appropriate valid methods are developed and applied.

In all research particular attention should be paid to co-ordination with other specialists, wherever possible.

There is a pressing need for the development of means of interpreting data obtained in experiments. Such interpretation should pay heed to the normal behaviour of microbial communities under what may be termed acceptable stress.

It is vital that those responsible for the introduc-

218

tion and imposition of regulatory requirements take full note of the advice of microbiologists and other specialists.

7. REFERENCES

Anon. (1981). OECD Guidelines for testing of chemicals. Section 304A. Inherent biodegradability in soil. OECD, Paris.

Domsch, K.H., Jagnow, G, Anderson, T.-H. (1983): An ecological concept for the assessment of side-effects of agrochemicals on soil micro-organisms. Residue Reviews 86, 65-105.

Greaves, M.P., Poole, N.J. Domsch, K.H., Jagnow, G. and Verstraete, W., (1980). Recommended tests for assessing the impact of pesticides on the soil microflora. Technical Report Agricultural Research Council Weed Research Organization, (59), 15 pp.

8. LIST OF PARTICIPANTS, 1978

Altman, J., c/o Institut für Pflanzenkrankheiten und Pflanzenschutz, Technische Universität Hannover, Herrenhäuser Str. 2, D 3000 Hannover, F.R. of Germany.

Anderson, J.P.E., Institut für Bodenbiologie, Bundesforschungsanstalt für Landwirtschaft, Bundesallee 50, D3300 Braunschweig, F.R. of Germany.

Anderson, T.H. Mrs, Institut für Bodenbiologie, Bundesforschungsanstalt für Landwirtschaft, Bundesallee 50, D 3300 Braunschweig, F.R. of Germany.

Andrews, J.H., Department of Plant Pathology, University of Wisconsin, 1630 Linden Drive, Madison, Wisc. 53706, U.S.A.

Bakalivanov, D., Institut po Pochvoznanie "N. Pushkarov", Sofia, Bulgaria.

Baumert, D., Institut für Pflanzenschutzforschung, Biologie, Schering AG, Gollanczstr. 71-99, D 1000 Berlin 28, F.R. of Germany.

Bollag, J.-M., Department of Agronomy, Pennsylvania State University, 119 Tyson Building, University Park, Penn.

16802, U.S.A.

Bollen, G.J., Laboratory of Phytopathology, Agric. University Wageningen, Binnenhaven 9, Wageningen, The Netherlands.

Bowen, G.D., Division of Soils, CSIRO, Glen Osmond, South Australia 5064, Australia.

Brunnert, H., Isotopenlaboratorium, Bundesforschungsanstalt für Landwirtschaft, Bundesallee 50, D 3300 Braunschweig, F.R. of Germany.

Cook, K.A., Shell Biosciences Laboratory, Sittingbourne Res. Centre, Sittingbourne, Kent ME9 8AG, England.

Deshmukh, V.A., Department of Soil Sci. and Agric. Chem. Punjabrao Krishi Vidyapeeth, Akola, Mararashtra 444104, India.

Dijk, H. van, Inst. voor Bodemvrachtbaarheid, Oosterweg 92, Haren, The Netherlands.

Doelman, P., Rijksinstituut voor Natuurbeheer, Kemperbergerweg 11, Arnhem, The Netherlands.

Domsch, K.H., Institut für Bodenbiologie, Bundesforschungsanstalt für Landwirtschaft, Bundesallee 50, D 3300 Braunschweig, F.R. of Germany.

Drew, E.A. Mrs, ICI Plant Protection Division, Jealott's Hill Research Station, Bracknell, Berks, England.

Egli, Th., Ciba-Geigy AG, R 1207, CH 4002 Basel, Switzerland.

Eichler, D., Celamerck, 6507 Ingelheim, F.R. of Germany.

Fischer, H., Institut für Bodenbiologie, Bundesforschungsanstalt für Landwirtschaft, Bundesallee 50, D 3300 Braunschweig, F.R. of Germany.

Fournier, J.C., Laboratoire de Microbiologie des Sols, 7 rue Sully, F 21034, Dijon Cedex, France.

Gijswijt, M.J., Philips Duphar BV, Agrobiol. Lab. "Boekesteyn", P.O. Box 4, 's Graveland, The Netherlands.

Greaves, M.P., Agric. Res. Council, Weed Res. Organization, Begbroke Hill, Sandy Lane, Yarnton, Oxford OX5 1PF, England.

Gunner, H.B., Department of Envir. Sciences, Marshall Hall, University of Massachusetts, Amherst, Mass. 01003, U.S.A.

Haider, K., Institut für Biochemie des Bodens, Bundesforschungsanstalt für Landwirtschaft, Bundesallee 50, D 3300 Braunschweig, F.R. of Germany.

Hermann, G., Bayer AG, Pflanzenschutz AT, Biologische Forschung, D 5090 Leverkusen, F.R. of Germany.

Heye, C.C., Plant Pathology Department, University of Wisconsin, Madison, Wisc. 53706, U.S.A.

Huge, P.L., Station de Chimie et de Physique Agric., 115 Ch. de Wavre, B 5800 Gembloux, Belgium.

Iloba, Ch., Department of Crop Science, University of Nigeria, Nsukka, Anambra State, Nigeria.

Jagnow, G., Institut für Bodenbiologie, Bundesforschungsanstalt für Landwirtschaft, Bundesallee 50, d 3300 Braunschweig, F.R. of Germany.

Jalali, B.L., Department of Plant Pathology, Haryana Agric. University, Hissar – 125004, Haryana, India.

Johnen, B.G., ICI Plant Protection Division, Fernhurst, Haslemere, Surrey, England.

Kaiser, P., Institut National Agronomique, 9 rue de l'Arbalete, F 75231 Paris Cedex 05, France.

Karanth, N., Institut für Bodenbiologie, Bundesforschungsanstalt für Landwirtschaft, Bundesallee 50, D 3300 Braunschweig, F.R. of Germany.

Kecskes, M., Res. Institut for Soil and Agric. Chem., Hungarian Acad. of Sciences, Herman Otto ut 15, H 1022 Budapest, Hungary.

Kelley, W.D., Botany and Microbiology Department, Auburn University, Auburn, Alabama, U.S.A.

Lebbink, G., Inst. voor Bodemvrachtbaarheid, Oosterweg 92, Haren-Gr., The Netherlands.

Lockwood, J.L., Department of Botany and Plant Pathology, Michigan State University, East Lansing, Mich. 48824, U.S.A.

Luscombe, B., May & Baker Ltd, Fyfield Road, Ongar, Essex, England.

Mahadevan, A., Centre of Advanced Study in Botany, University of Madras, University Buildings, Madras - 5, India.

Malkomes, H.-P., Institut für Unkrautforschung, Biologische Bundesanstalt, Messeweg 11/12, D 3300 Braunschweig, F.R. of Germany.

Martens, R., Institut für Bodenbiologie, Bundesforschungsanstalt für Landwirtschaft, Bundesallee 50, D 3300 Braunschweig, F.R. of Germany.

Moubasher, A.H., Faculty of Science, University of Assiut, Egypt.

Nilsson, H.E., Department of Pl. and Forest Protection, Swedish Univ. of Agric. Science, S 75007 Uppsala 7, Sweden.

Nowak, A., c/o Universität Hohenheim, Institut für Phytomedizin, D 7000 Stuttgart-Hohenheim, F.R. of Germany.

Papavizas, G.C., Soilborne Disease Lab., Plant Protection Institute, USDA, Beltsville, Maryland 20705, U.S.A.

Poole, N.J., Plant Protection Division, Jealott's Hill Research Station, Bracknell, Berks, England.

Pugh, G.J.F., Department of Biological Sciences, University of Aston, Birmingham B4 7ET, England.

Reber, H., Institut für Bodenbiologie, Bundesforschungsanstalt für Landwirtschaft, Bundesallee 50, D 3300 Braunschweig, F.R. of Germany.

Rodriguez-Kabana, R., Botany and Microbiology Department, Auburn University, Auburn, Alabama 36830, U.S.A.

Rovira, A.D., Division of Soils, CSIRO, Glen Osmond,

South Australia 5064, Australia.

Sumner, D.R., Plant Pathology Department, Coastal Plain Exp. Station, Tifton, Georgia 31794, U.S.A.

Taubel, N., Hoechst AG, Landw. Entw. Abteilung, Hessendamm 1-3, D 6234 Hattersheim 1, F.R. of Germany.

Torstensson, L., Department of Microbiology, The Swedish University of Agric. Sciences, S 75007 Uppsala 7, Sweden.

Verstraete, W., Lab. Microbial Ecology, University of Gent, Coupure L653, B 9000 Gent, Belgium.

Vlassak, K., Laboratory of Soil Fertility and -Biology, Kard. Mercierlaan 92, B 3030 Leuven, Belgium.

Vonder Mühll, F., Dr. R. Maag AG, CH 8157 Dielsdorf, Switzerland.

Wallnofer, P., Bayerische Landesanstalt für Bodenkulture und Pflanzenbau, Abt. Pflanzenschutz, Menzinger Str. 54, D 8000 München 19, F.R. of Germany.

Wiedemann, A. Mrs, Institut für Pflanzenpathologie, Griesebachstr. 6, 3400 Göttingen, F.R. of Germany.

Ziedan, M., Institut of Plant Pathology, Agric. Res. Center No. 5-A, Giza, Egypt.

LIST OF PARTICIPANTS, 1979

Adcock, John, Fisons Ltd, Chesterford Park Research Station, nr Saffron Walden, Essex, UK.

Anderson, Heidi, Institut für Bodenbiologie, Bundesforschungsanstalt für Landwirtschaft, Bundesallee 50, D 3300 Braunschweig, F.R. of Germany.

Anderson, John P.E., Institut für Bodenbiologie, Bundesforschungsanstalt für Landwirtschaft, Bundesallee 50, D 3300 Braunschweig, F.R. of Germany.

Burns, Richard, University of Kent, Canterbury, Kent, UK.

Cook, Keith, Shell Research Centre, Sittingbourne, Kent, UK.

Debourge, Jean-Claude, Rhone-Poulenc Phytosanitaire, 15-21 rue Pierre Baizet, 69009 Lyon, France.

Dijk, Henrik van, Inst. voor Bodemvrachtbaarheid Oosterweg 92, Haren, The Netherlands.

Doelman, Peter, Research Institute of Nature Management, Kemperbergerweg 67, Arnhem, The Netherlands.

Domsch, K.H., Institut für Bodenbiologie, Bundesforschungsanstalt für Landwirtschaft, Bundesallee 50, D 3300 Braunschweig, F.R. of Germany.

Drew, Elizabeth, ICI, Plant Protection Division, Jealott's Hill Research Station, Bracknell, Berks RG12 6EY, UK.

Eichler, Dietrich, Celamerck GmbH & CokG, D 6507 Ingelheim, F.R. of Germany.

Ellgehausen, Holm, Ciba-Geigy Ltd, CH 4002 Basel, Switzerland.

Gijswijt, Theo, Philips Duphar BV, Agrobiol. Lab. "Boekesteyn", P.O. Box 4, 's Graveland, The Netherlands.

Greaves, Mike, Agric. Res. Council, Weed Res. Organization, Begbroke Hill, Yarnton, Oxford OX5 1PF, UK.

Haanstiam, L., Research Institute of Nature Management, Kemperbergerweg 67, Arnhem, The Netherlands.

Harris, Peter, University of Reading, Reading, Berks, UK.

Herman, Gunther, Bayer AG, Pflanzenschutz AT, Biologische Forschung, D 5090 Leverkusen, F.R. of Germany.

Iwan, Jorg., Schering AG, Postfach 65 03 11, D 1000 Berlin 65, F.R. of Germany.

Jagnow, Gerhard, Institut für Bodenbiologie, Bundesforschungsanstalt für Landwirtschaft, Bundesallee 50, D 3300 Braunschweig, F.R. of Germany.

Johnen, Bernhard, ICI, Plant Protection Division, Fernhurst, Haslemere, Surrey, UK.

Kelley, Walter, Botany and Microbiology Department, Auburn University, Auburn, Alabama, U.S.A.

Lebbink, Gerrit, Inst. voor Bodemvrachtbaarheid, Oosterweg 92, Haren–Gr., The Netherlands.

Luscombe, Brian, May & Baker Ltd, Fyfield Road, Ongar, Essex, UK.

Malkomes, H.-P., Institut für Unkrautforschung, Biologische Bundesanstalt, Messeweg 11/12, D 3300 Braunschweig, F.R. of Germany.

Osborne, Gary, Agricultural Research Institute, Wagga–Wagga, New South Wales, Australia.

Poole, Nigel, ICI, Plant Protection Division, Jealott's Hill Research Station, Bracknell, Berks., RG12 6EY, U.K.

Rodrigues–Kabana, R. Botany and Microbiology Department, Auburn University, Auburn, Alabama 36830, U.S.A.

Samir, Salem, Faculty of Agriculture, Zagazig Univeristy, Zagazig, Egypt.

Somerville, L., FBC Limited, Chesterford Park Research Station, Saffron Walden, Essex, CB10 1XL, U.K.

Spier, Tom, DSIR Soil Bureau, New Zealand.

Taubel, Norbert, Hoechst AG, Lanw. Entw. Abteilung, Hessendamm 1-3, D 6234 Hattersheim 1, F.R. of Germany.

Torstensson, Lennart, Department of Microbiology, The Swedish University of Agric. Sciences, S 75007 Uppsala 7, Sweden.

Vaagt, Gero, Seeblick 3, 2300 Kiel 1, F.R. of Germany.

Varner, Reed, Du Pont Experimental Station, Wilmington, Delaware 19898, U.S.A.

Verstraete, Willy, Lab. Microbial Ecology, University of Gent, Coupure L653, B 9000 Gent, Belgium.

Vonder Mühll, F., Dr. R. Maag AG, CH 8157 Dielsdorf, Switzerland.

Wiedemann, Almuth, Institut für Pflanzenpathologie, Griesebachst. 6, 3400 Göttingen, F.R. of Germany.

LIST OF PARTICIPANTS, 1985

Anderson, J.P.E., Bayer AG, PF-F/CE, Institut für Ökobiologie, Zentrum Landwirtschaft, Monhein, D 5090 Leverkusen, F.R. of Germany.

Arnold, D.J., Metabolism Department, Schering Agrochemicals Limited, Chesterford Park Research Station, Saffron Walden, Essex CB10 1XL, UK.

Bagnall, B.H., Bayer UK Ltd, Agrochem. Division, Eastern Way, Bury St. Edmunds, Suffolk IP32 7AH, UK.

Brümmer, Dept Plant Nutr. and Soil Sci., University of Kiel, Olshausenstr. 40-60, D 23 Kiel, F.R. of Germany.

Burns, R., Biological Laboratory, University of Kent, Canterbury, Kent CT2 7NJ, UK.

Caley, V.J., Dept of Plant Protection, Queensland Agricultural College, Lawes (Gatton), Queensland 4343, Australia.

Castle, D., ICI Plant Protection Div., Jealott's Hill Research Station, Bracknell, Berks RG12 6EY, UK.

Cook, K.A., Shell Research Ltd, Sittingborne Research Centre, Sittingborne, Kent ME9 8AG, UK.

Dijk, H. van, Inst. for Soil Fertility, PO Box 3003, 9750 RA Haren Gn, The Netherlands.

Domsch, K.H., Institut für Bodenbiologie, Bundesforschungsanstalt für Landwirtschaft, Bundesallee 50, D 3300 Braunschweig, F.R. of Germany.

Doorn, A.M. van, Institute for Pesticide Research, Maykeweg 22, 6709 Pg Wageningen, The Netherlands.

Eichler, D., Celamerck GmbH & CokG, D 6507 Ingelheim, F.R. of Germany.

Ellgehausen, H., Research & Consulting Co. AG, Zelgliweg 1, CH 4452 Itingen/BL, Switzerland.

Elmholt, S., Statens Planteavlsforsog, Institut for Ukrudsbekaempelse, Flakkebjerg, DK 4200 Salgelse, Denmark.

Fischer, H., Umweltbundesamt, Bismarckplatz 1, 1 Berlin 33, Germany.

Fournier, J.-C., INRA, 17 rue Sully, 21023 Dijon Cedex, France.

Gestel, C.A.M. van, RIVM, PO Box 1, 3720 BA Bilthoven, The Netherlands.

Greaves, M.P., Long Ashton Research Station, Weed Research Division, Long Ashton, Bristol BS18 9AF, UK.

Goksoyr, J., Dept of Microbiology and Plant Physiology, Allegaten 70, N 5000 Bergen, Norway.

Halley, B.A., Merck, Sharp and Dohme Research Laboratories, PO Box 2000, Rahway, N. Jersey 07065, U.S.A.

Hamm, R.T., BASF AG, Land. Versuchsstation, Postfach 220, D 06703 Limburgerhof, F.R. of Germany.

Herman, G., Bayer AG, PF-F/CE, Institut für Okbiologie, Zentrum Landwirtschaft, D 5090 Leverkusen, F.R. of Germany.

Horemans, S., Laboratory of Soil Fertility and Biology, Kard, Mercierlaan 92, B 3030 Leuven, Belgium.

Hossack, D.J.N., Head of Analytical Chemistry and Microbiology, Huntingdon Res. Centre, Huntingdon, Cambridgeshire PE18 6ES, UK.

Hudson, G.H., Head of Divisional Regulations Relating to Crop Protection and Animal Husbandry, Dir. General for Agriculture, 200 rue de la Loi, 1049 Brussels, Belgium.

Johnson, B., University of North Wales, Dept of Biochemistry and Soil Science, Memorial Buildings, Deniniol Road, Bangor, Gwynedd, UK.

Leake, C.R., Metabolism Department, Shcering Agrochemicals Limited, Chesterford Park Research Station, Saffron Walden, Essex CB10 1XL, UK.

Luotola, M., Senior Research Officer, Ministry of the Environment, PO Box 306, SF 00531 Helsinki 53, Finland.

Luscombe, B., May & Baker Ltd, Fyfield Road, Ongar, Essex, UK.

Malkomes, H.-P., Biologische Bundesanstalt für Land- und Forstwirtschaft, Messeweg 11/12, D 3300 Braunschweig, F.R. of Germany.

Marsh, J.A.P., Charlbury Environmental Consultancy, 6 The Slade, Charlbury, Oxfordshire, UK.

Morgensen, B.B., Nat. Agency of Environ. Protection Station, Dept of Pesticides, Strangade 29, DK 1401 Copenhagen K, Denmark.

Newby, S.E., Metabolism Dept, Schering Agrochemicals Limited, Chesterford Park Research Station, Saffron Walden, Essex CH10 1OX, UK.

Plueckem, U., Ciba Geigy Ltd, R 1061.2.13, CH 4002 Basel, Switzerland.

Poole, N., ICI Plant Protection Division, Jealott's Hill Res. Station, Bracknell, Berkshire RG12 6EY, UK.

Prinold, S., Shell Research Ltd, Sittingborne Research Centre, Sittingborne, Kent ME9 8AG, UK.

Schuepp, H., Swiss Federal Research Stn for Fruit Growing, Viticulture and Horticulture, CH 8820 Wadenswil, Switzerland.

Somerville, L., Environmental Sciences Department, Schering Agrochemicals Limited, Chesterford Park Research Station, Saffron Walden, Essex, CB10 1XL, UK.

Soulas, G., INRA, 27 rue Sully, Dijon Cedex, France.

Taubel, N., Hoechst AG, D 6230 Frankfurt 80, F.R. of Germany.

Torstensson, L., Dept of Microbiology, Box 7025, The Swedish University of Agric. Sci., S 75007 Uppsala 7, Sweden.

Torsvik, V., Dept of Microbiology, Allegaten 70, N 5000 Bergen, Norway.

Venkatramesh, M., Rallis Agrochemical Research, Plot Nos 21 and 22, Phase 2, Peenya Industrial Area, PO Box 5813, Bangalore 560 058, India.

Verstraete, W., Lab. Microbial Ecology, University of Gent, Coupure L653, B 9000 Gent, Belgium.

Vonder Mühll, P.A., Dr. R. Maag AGm, CH 8157 Dielsdorf, Switzerland.

Vonk, J., TNO, PO Box 108, 3700 AC Zeist, The Netherlands.

Werf, H. van der, Lab. voor Alg. en Ind. Microbiologie, Fac. landb. Wetensch, R.U.G. Coupure 533, B 9000 Gent, Belgium.

Westlake, G., MAFF, Tolworth Laboratory, Hook Rise South, Tolworth, Surbiton, Surrey, UK.

Willems, T., Duphar BV Crop Protection Division, PO Box 4, 1243 ZG S-Graveland, The Netherlands.

Wilson, A., Dept of the Environment, Room A342, Romney House, 43 Marsham Street, London SW1P 3PY, UK.

Wisson, M., Agro-Development Analytical Laboratory, Sandoz, Lichstrasse, CH 4000 Basel, Switzerland.

Wynn-Williams, D., British Antarctic Survey, Madingley Road, Cambridge CB3 OET, UK.